Analytical Profiles

of

Drug Substances

and

Excipients

EDITORIAL BOARD

Analytical Profiles of Drug Substances and Excipients

Volume 28

edited by

Harry G. Brittain

Center for Pharmaceutical Physics
10 Charles Road
Milford, New Jersey 08848

Founding Editor:

Klaus Florey

San Diego San Francisco New York Boston London Sydney Tokyo

Academic Press Rapid Manuscript Reproduction

This book is printed on acid-free paper. 

Academic Press
A Harcourt Science and Technology Company
525 B Street, Suite 1900, San Diego, California 92101-4495, USA
http://www.academicpress.com

Academic Press
Harcourt Place, 32 Jamestown Road, London NW1 7BY, UK
http://www.academicpress.com

International Standard Book Number: 0-12-260828-3

PRINTED IN THE UNITED STATES OF AMERICA
01 02 03 04 05 06 SB 9 8 7 6 5 4 3 2 1

CONTENTS

AFFILIATIONS OF EDITORS AND CONTRIBUTORS

Abdullah A. Al-Badr: Department of Pharmaceutical Chemistry, College of Pharmacy, King Saud University, P.O. Box 2457, Riyadh-11451, Saudi Arabia

J. Al-Zehouri: Department of Pharmaceutical Chemistry, College of Pharmacy, King Saud University, P.O. Box 2457, Riyadh-11451 Saudi Arabia

Harry G. Brittain: Center for Pharmaceutical Physics, 10 Charles Road, Milford, NJ 08848-1930, USA

Alekha K. Dash: Department of Pharmaceutical & Administrative Sciences, School of Pharmacy and Allied Health Professions, Creighton University, Omaha, NE 68178, USA

Humeida A. El-Obeid: Department of Pharmaceutical Chemistry, College of Pharmacy, King Saud University P.O. Box 2457, Riyadh-11451 Saudi Arabia

H.I. El-Subbagh: Department of Pharmaceutical Chemistry, College of Pharmacy, King Saud University, P.O. Box 2457, Riyadh-11451 Saudi Arabia

Klaus Florey*:* 151 Loomis Court, Princeton, NJ 08540, USA

Yuri Goldberg: Apotex Inc., 150 Signet Drive, Toronto, Ontario M9L 1T9 Canada

Kimio Higashiyama: Institute of Medicinal Chemistry, Hoshi University, 4-41, Ebara 2-chome, Shinagawa-ku, Tokyo 142-8501, Japan

Gunawan Indrayanto*:* Laboratory of Pharmaceutical Biotechnology, Faculty of Pharmacy, Airlangga University, Jl. Dharmawangsa dalam, Surabaya 60286, Indonesia

Dominic P. Ip*:* Merck, Sharp, and Dohme, Building 78-210, West Point, PA 19486, USA

Vijay K. Kapoor: University Institute of Pharmaceutical Sciences, Panjab University, Chandigarh 160 014 India

Hajime Kubo: Institute of Medicinal Chemistry, Hoshi University, 4-41, Ebara 2-chome, Shinagawa-ku, Tokyo 142-8501, Japan

Krishan Kumar: Merial Limited, 2100 Ronson Road, Iselin, NJ 08830, USA

David J. Mazzo: Preclinical Development, Hoechst Marion Roussel, Inc., Route 202-206, P.O. Box 6800, Bridgewater, NJ 08807, USA

Mohammed S. Mian: Department of Pharmaceutical Chemistry, College of Pharmacy, King Saud University P.O. Box 2457, Riyadh-11451 Saudi Arabia

Achmad Toto Poernomo: Faculty of Pharmacy, Airlangga University, Jl. Dharmawangsa Dalam, Surabaya 60286, Indonesia

Hadi Poerwono: Faculty of Pharmacy, Airlangga University, Jl. Dharmawangsa Dalam, Surabaya 60286, Indonesia

Leroy Shervington: Pharmacy Faculty, Applied Science University, Amman 11931, Jordan

Amarjit Singh: Research & Development Centre, Panacea Biotec Ltd., P.O. Lalru 140 501 India

Paramjeet Singh: Research & Development Centre, Panacea Biotec Ltd., P.O. Lalru 140 501 India

I. Ketut Sudiana: Faculty of Medicine, Airlangga University, Jl. Mayjen Prof. Dr. Moestopo 47, Surabaya 60131, Indonesia

Suharjono: Faculty of Pharmacy, Airlangga University, Jl. Dharmawangsa Dalam, Surabaya 60286, Indonesia

Timothy J. Wozniak: Eli Lilly and Company, Lilly Corporate Center, MC-769, Indianapolis, IN 46285, USA

PREFACE

The comprehensive profiling of drug substances and pharmaceutical excipients as to their physical and analytical characteristics continues to be an essential feature of drug development. The compilation and publication of comprehensive summaries of physical and chemical data, analytical methods, routes of compound preparation, degradation pathways, uses and applications, *etc.*, has always been a vital function to both academia and industry. Historically, the *Analytical Profiles* series has always provided this type of information, and of the highest quality available.

However, the need for information continues to grow along many new fronts, and the appropriate mechanisms to summarize what workers in the pharmaceutical field need to know must develop accordingly. It has therefore become time to re-consider the current format of Profile chapters, and to re-define these so as to be more useful for the community at large. In the past, a typical chapter essentially consisted of a summary of the physical characteristics and analytical methods for either drug substances or excipients, and in some notable instances has also contained extensive information regarding other aspects of the compound in question.

In the future, Profile chapters will be more focused on a particular aspect of a drug substance or an excipient. Profile chapters will therefore fall into one of the following main categories:

1. Physical characterization of a drug substance or excipient
2. Analytical methods for a drug substance or excipient
3. Detailed discussions of the clinical uses, pharmacology, pharmacokinetics, safety, or toxicity of a drug substance or excipient
4. Reviews of methodology useful for the characterization of drug substances or excipients

More information about the new content will be forthcoming, as will a name-change for the series. I will welcome communications from anyone in the pharmaceutical community who might want to provide an opinion.

Harry G. Brittain

CITRIC ACID

Hadi Poerwono[1], Kimio Higashiyama[3], Hajime Kubo[3], Achmad Toto Poernomo[1], Suharjono[1], I Ketut Sudiana[2], Gunawan Indrayanto[1], and Harry G. Brittain[4]

(1) Faculty of Pharmacy, Airlangga University
Jl. Dharmawangsa Dalam, Surabaya 60286
Indonesia

(2) Faculty of Medicine, Airlangga University
Jl. Mayjen Prof. Dr. Moestopo 47, Surabaya 60131
Indonesia

(3) Institute of Medicinal Chemistry, Hoshi University
4-41, Ebara 2-chome, Shinagawa-ku, Tokyo 142-8501
Japan

(4) Center for Pharmaceutical Physics
10 Charles Road, Milford, NJ 08848
USA

1075-6280/01 $35.00

Contents

1. Description

1.1 Nomenclature

1.1.1 Systematic Chemical Name

2-Hydroxypropane-1,2,3-tricarboxylic acid

1.1.2 Nonproprietary Names

1.1.2.1 Anhydrate

Acidum Citricum Anhydricum; Anhydrous Citric Acid; Citronensäure; E330; β-Hydroxytricarballylic Acid [1-5]

1.1.2.2 Monohydrate

Acido del Limón; Acidum Citricum Monohydricum; Citric Acid Monohydrate; Hydrous Citric Acid [1,4]

1.1.3 Proprietary Names

1.1.3.1 Anhydrate

Uriflex (FM) [1]

1.1.3.2 Monohydrate

Citrosteril; Paediatric Compound Tolu Linctus BP 1999; Paediatric Simple Linctus BP 1999; Simple Linctus BP 1999 [1]

1.2 Formulae

1.2.1 Empirical Formula, Molecular Weight, CAS Number

1.1.3.1 Anhydrate

$C_6H_8O_7$ [MW = 192.1) CAS number = 72-99-2

1.1.3.2 Monohydrate

$C_6H_8O_7 \cdot H_2O$ [MW = 210.1] CAS number = 5949-29-1

1.2.2 Structural Formula

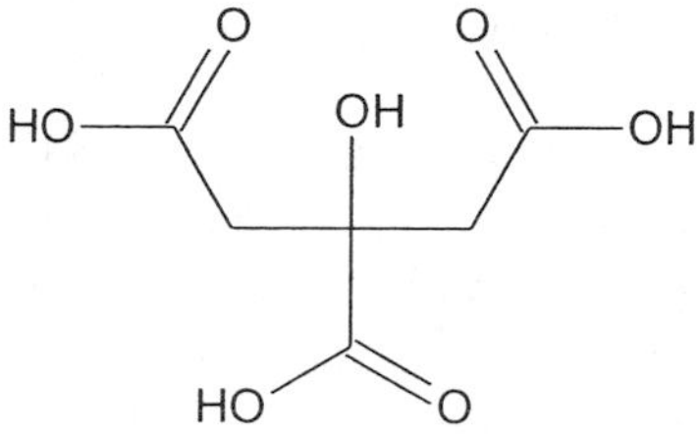

1.3 Elemental Analysis

The calculated elemental composition is as follows [2]:

carbon:	37.51%
hydrogen:	4.20%
oxygen:	58.29%

1.4 Appearance

1.1.3.1 Anhydrate

Odorless or almost odorless, colorless crystals or granules, or a white crystalline powder [1,2,4,6-12].

1.1.3.2 Monohydrate

Odorless or almost odorless, colorless and translucent crystals or granules or white crystalline powder. Pleasant, sour taste [1,2,4,6-12].

1.5 Uses and Applications

Citric acid occurs in the terminal oxidative metabolic system of all but a very few organisms. This system, variously referred to as the Krebs cycle, the tricarboxylic acid cycle, or the citric acid cycle, is a metabolic intermediate cycle involving the terminal steps in the conversion of carbohydrates, fats, or proteins to carbon dioxide and water with concomitant release of energy necessary for growth, movement, luminescence, chemosynthesis, and reproduction. The cycle also provides the carbonaceous materials from which amino acids and fats are synthesized by the cell [6,12-14].

Citric acid chelates calcium and is used by blood banks, since the citrate-calcium chelate helps to prevent blood coagulation. Citrate was found to be preferable to heparin as an anticoagulant during hemodialysis with cellulose triacetate dialyzers. Presumably by its chelation of calcium ions, citrate treatment reduced the release of indicators of degranulation of polymorphonuclear cells. Because citric acid is a very efficient binder of calcium, it has been used in chelation therapy for heavy-metal contamination [1,2,9,15,16,79].

A specific effect of citric acid treatment on isolated cell nuclei from bovine cerebral cortex results in extraction of a specific set of proteins from the nuclei rather than separation of inner and outer nuclear membranes [17]. Use of a citric acid aerosol spray has been reported as a safe, convenient, and efficacious smoking cessation aid. Citric acid can be substituted for acetic acid as the acidifying agent in dialysates. Because citric acid is a dry powder, its use allows for formulation of dry dialysates, which offer convenience for transportation and storage [79].

Citric acid is used in effervescing mixtures, and citric acid monohydrate is used in the preparation of effervescent granules. Citric acid monohydrate is used as a synergist to enhance the effectiveness of antioxidants [1,2,4,6]. A combination of citric acid and glycine was used to improve the dissolution stability of hard gelatin capsules by preventing pellicle formation or cross-linking of the capsule gelatin [18]. The use of co-formulated citric acid and low substituted hydroxypropyl-cellulose results in extremely fast dissolution of carbamazepine from solid dosage forms [19].

Citric acid is used in beverages, jellies, jams, preserves, and candy to provide tartness, acid taste. It is used in the manufacture of alkyd resins, in its esterified form as a plasticizer and foam inhibitor, in manufacture of citric acid salts, an aid to facilitate abscission of fruit in harvesting, and in cultured dairy products. Citric acid has also been used to adjust the pH of foods and cosmetics, as a sequestering agent to remove trace metals, and as a mordant to brighten colors. It has been used in electroplating, in special inks, in analytical chemistry for determining citrate-soluble P_2O_5, and as reagent for albumin, mucin, glucose, and bile pigments [2,4-6].

Preparations containing citric acid are used in the management of dry mouth and to dissolve urinary bladder calculi, alkalinize the urine, and prevent encrustation of urinary catheters. Citric acid is an ingredient of citrated anticoagulant solutions; and also a reactive intermediate in chemical synthesis. Citric acid has also been used in preparations for the treatment of gastrointestinal disturbances and chronic metabolic acidosis caused by chronic renal insufficiency or syndrome of renal tubular acidosis [1,2,4-6,20,21,79].

Lemonade therapy was effective in elevating urine citrate levels in 12 patients with hypocitraturic calcium nephrolithiasis. Lemons contain a high level of citric acid (49.2 g/kg) as compared with other citrus fruits. The lemonade treatment resulted in good patient compliance, and was well tolerated and inexpensive. Direct infusion of a citrate solution, which was prepared from ten grams of citrate powder (2.0 g of citric acid, 4.0 g of monopotassium citrate, and 4.0 g of monosodium citrate) in 100 mL of distilled water, into the pancreatic ducts dissolved calcifications in two women with chronic abdominal pain caused by pancreatic duct obstruction [79].

In Great Britain, citric acid (1 in 500 parts of water) is an approved disinfectant for foot and mouth disease [1]. Citric acid is well known as a disinfecting agent used for dialysis equipment. Dialyzer reprocessing by using 1.5% citric acid heated to 95°C for 20 hours is an alternative method that produces equivalent microbiologic effects [22]. Nagoba *et al* reported an effective and economical approach for the treatment of superficial *pseudomonas* infections with citric acid [23].

Citric acid is a commodity chemical produced and consumed throughout the world. It is reported that the global citric acid demand is somewhere between 950,000 and 975,000 metric tons per year. Global capacity in 1998 was almost 2 billion pounds (880 thousand metric tons). Capacity utilization in two largest producing regions was 88% by the United States and 85% by Western Europe. Approximately two-thirds of the citric acid produced is consumed in the food and beverage industry, primarily as an acidulant [24-26].

The citric acid industry is highly concentrated and very competitive. In 1999, five producing companies accounted for 65% of the world's total citric acid capacity, and nearly 90% of the world's refined and purified citric acid capacity [24].

2. Method(s) of Preparation

Citric acid is the main acid present in lemons, oranges, pineapples, strawberries, red currants, cranberries, and other fruit. For this reason the traditional method of preparing citric acid was by extraction from the juice of certain citrus fruit, such as lemons (contains about 5-10% of citric acid) and limes, and later from pineapple wastes.

With the development of fermentation technology and chemical synthesis, the traditional processes of citric acid preparation by juice extraction became obsolete.

Citric acid is found in milk, and in small amounts in blood and in urine. The substance plays a vital role as an intermediate in metabolic processes [2-4,6,27-29]. The major pathway for the formation of citric acid is through the condensation of oxaloacetate and acetyl-CoA catalyzed by citrate synthase, the first enzyme in the tricarboxylic acid (TCA) cycle. In the glyoxylate cycle, a modified form of the TCA cycle, citric acid is also formed from acetyl-CoA arising from fatty acid breakdown [6,12-14].

2.1 Fermentation

The general manufacture of citric acid is by the fermentation of solutions of glucose, sucrose, or purified cane-molasses in the presence of certain inorganic salts and air, by various molds or fungi, such as *Citromyces pfefferianus*, *Aspergilus niger*, *Penicillium sp* [2-4,6,29-38].

Domestic citric acid supplies in United States are ample owing to an abundance of feedstock corn, which is the main source of dextrose used to produce citric acid. Other organic sources may be used to provide dextrose, such as cane sugar and wheat, but corn is the most important. Corn accounts for almost all citric acid produced in Unites States [39]. China's major citric acid producer has a goal to process 1.3 million tons of corn by the end of 2001. This capability will produce yield by fermentation of minute organisms with the use of corn powder, which include organic acids, amino acids, and glutamates [37].

A number of fermentation processes are available for the preparation of citric acid in commercial quantities.

2.1.1 Surface Process

The surface methodology consists of inoculating with spores of *A. niger* shallow aluminum or stainless steel pans containing sugar solution along with sources of accessible nitrogen, phosphate, magnesium, and various trace minerals. Growth of the mold occurs on the solution surface forming a rubbery, convoluted mycelial mass. Air passed over the surface provides oxygen, and controls the temperature by evaporative cooling.

Because of its relatively low cost, molasses has proven to be a preferred source of sugar for microbial production. Since it is a by-product of sugar refining, the content of molasses varies considerably and not all types are suitable for citric acid production. Beet molasses is preferred to cane, but there are considerable yield variations within each type. Beet molasses must be treated prior to use so as to improve yield. Treatment may consist of ion-exchange, chemical precipitation, or chelation of metallic ions [6].

2.1.2 Submerged Process using *Aspergillus niger*

In this process the microorganism (*A. niger*) is grown dispersed through a liquid medium. The fermentation vessel usually consists of a sterilizable tank having a capacity of several hundred cubic meters (thousands of gallons) equipped with a mechanical agitator and a means of introducing sterile air.

In practice, the mold spores are produced under controlled aseptic conditions. When harvested they are used in specific quantities to seed the inoculum fermenter, which is a stage prior to the production fermenter that contains a medium designed to develop cellular mass and to control morphology rather than producing acid. The inoculum is then transferred aseptically to the production fermenter. Medium constituents in this stage are chosen to foster acid production rather than growth. Progress is monitored by periodically determining the acid and sugar content of the vessel. In addition, pH, dissolved oxygen, and solids content are recorded.

Initial sugar concentrations and acid yield compare favorably with those of the surface process. As in the surface process, molasses requires treatment prior to use. Selection of raw materials for the submerged process suffers from the same constraints as the surface method, except that it permits a wider choice of materials [6].

The possible use of concentrated milk-wastewater as a fermentation medium for the production of citric acid by *A. niger* ATCC-9142 has been investigated. The addition of Mn(II), Fe(II), and Cu(II) to the medium promoted citric acid production, while only Mg(II) decreased citric acid production. This is supported by a review describing that the production of organic acids by fungi has profound implications for metal speciation, physiology, and biogeochemical cycles. Citric acid is an intermediate in the tricarboxylic acid cycle, with metals greatly influencing biosynthesis. Growth limiting concentration of Mn(II), Fe(II), and Zn(II) are important for high yields [31,33].

For economic production of citric acid, the possible use of a rarely used agro-industrial by product, maize starch-hydrolysate, has been explored. Screening of seventeen strains of *A. niger* for their capacity to produce citric acid using starch-hydrolysate as a substrate led to the selection of ITCC-605 as the most efficient strain for further improvement in citric acid content by mutation. The optimized culture conditions were concentration of starch-hydrolysate, 15% (glucose equivalent); ammonium nitrate, 0.25%; KH_2PO_4, 0.15%; nicotinic acid, 0.0001%, and initial pH of 2.0. Under these conditions, the mutant strain UE-1 yielded 490 g citric acid per kg of glucose consumed in 8 days of incubation at 30°C [32].

Citric acid production from xylan and xylan hydrolysate was effected by *A. niger* Yang no. 2 cultivated in a semi-solid culture using bagasse as a carrier. Yang no.2 produced 72.4 g/L and 52.6 g/L of citric acid in 5 days from 140 g/L of xylose and arabinose, respectively. Yang no. 2 produced 51.5 g/L of citric acid in 3 days from a concentrated xylan hydrolysate prepared by cellulase treatment, containing 100 g/L of reducing sugars. Moreover, Yang no. 2 directly produced 39.6 g/L of citric acid maximally in 3 days from 140 g/L of xylan [34].

2.1.3 Submerged Process using Yeast

A methodology issued in 1970 demonstrated the production of citric acid by species of yeast, e.g., *Candida guilliermondii*, grown submerged in a medium containing either glucose or blackstrap molasses with an equivalent amount of sugar. Fermentation time was shorter than with *A. niger*.

Candida strains are also used in a novel process that permits production of citric acid from hydrocarbons. In 1970 a procedure was issued for the conversion of C_9 to C_{20} normal paraffins by *Candida lipolytica*. Other methods issued in 1972 describe a procedure for selecting Candida mutans. One example shows a more than 100% citric acid weight yield from a C_{13} to C_{15} *n*-paraffin mixture. In 1974 Pfizer issued a continuous process for fermentation by *C. lipolytica* using a single vessel to which paraffin is continuously added and fermented broth continuously withdrawn [6].

2.2 Chemical Syntheses

A number of syntheses of citric acid have appeared in the chemical and patent literature since the 1880 report by Grimoux and Adam of a route based on the reaction of glycerol-derived 1,3-dichloroacetone with cyanide [3]. These synthetic methods include a number of recent syntheses aimed at discovering routes competitive with fermentation [6].

$$\begin{matrix} CH_2OH \\ | \\ CHOH \\ | \\ CH_2OH \\ \text{Glycerol} \end{matrix} \xrightarrow{HCl} \begin{matrix} CH_2Cl \\ | \\ CHOH \\ | \\ CH_2Cl \end{matrix} \xrightarrow{[O]} \begin{matrix} CH_2Cl \\ | \\ CO \\ | \\ CH_2Cl \\ \text{1,3-Dichloroacetone} \end{matrix} \xrightarrow{HCN} \begin{matrix} CH_2Cl \\ | \\ CH(OH)CN \\ | \\ CH_2Cl \end{matrix} \xrightarrow{KCN} \begin{matrix} CH_2CN \\ | \\ CH(OH)CN \\ | \\ CH_2CN \end{matrix} \xrightarrow{H^+} \begin{matrix} CH_2CO_2H \\ | \\ CH(OH)CO_2H \\ | \\ CH_2CO_2H \\ \text{Citric acid} \end{matrix}$$

2.2.1 Conversion of acetonedicarboxylate and derivatives to the cyanohydrins followed by hydrolysis.

2.2.2 Reformatsky and Grignard reactions on oxaloacetic and glyoxylic esters [3].

$$\begin{matrix} CH_2Br \\ | \\ CO_2C_2H_5 \end{matrix} + \begin{matrix} COCO_2C_2H_5 \\ | \\ CH_2CO_2C_2H_5 \end{matrix} + Zn \longrightarrow C_2H_5O_2CCH_2\overset{\overset{OZnBr}{|}}{\underset{\underset{CH_2CO_2C_2H_5}{|}}{C}}CO_2C_2H_5 \xrightarrow{acid} HO_2CCH_2\overset{\overset{OH}{|}}{\underset{\underset{CH_2CO_2H}{|}}{C}}CO_2H$$

Citric acid

2.2.3 Photolysis of a mixture of glycolic and malic esters.

2.2.4 Benzilic acid rearrangement of 3,4-diketoadipic acid.

2.2.5 Oxidative degradation of quinic acid.

2.2.6 Aldol condensation of pyruvate with oxaloacetic acid to form citroylformic acid and lactone followed by oxidative decarboxylation.

2.2.7 Base-catalyzed condensation of acetonedicarboxylate with formate.

2.2.8 Conversion of γ-chloroacetoacetic acid to the cyanohydrin followed by hydrolysis, displacement of chloride by cyanide, and hydrolysis.

2.2.9 Electrolytic oxidation and then reduction of succinic acid to form a mixture of acids including citric acid.

The early syntheses above were followed by the more recent diverse synthetic approaches to citric acid from relatively inexpensive starting materials as described in the patent literature [6].

2.2.10 Oxidative conversion of maleic and fumaric derivatives to the corresponding oxaloacetates followed by cycloaddition with ketene to form β-lactones and then hydrolysis.

2.2.11. Bimolecular decarboxylative condensation of oxaloacetic acid to form citroylformic acid followed by oxidative decarboxylation.

2.2.12 Carboxylation of acetone using an alkali metal phenolate catalyst in solvents such as dimethylformamide or glyme (dimethyl ether and ethylene glycol) to form acetonedicarboxylate, followed by reaction with cyanide to form the cyanohydrin, and then acid or base hydrolysis.

2.2.13 Condensation of ketene and phosgene to form acetonedicarboxylic acid chloride or its cyclic precursor, reaction with alcohol to form the corresponding acetonedicarboxylate, followed by cyanide reaction to form cyanohydrin, and then acid hydrolysis.

2.2.14 Chlorination of diketene and hydrolysis to form 4-chloroacetoacetic acid, conversion to cyanohydrin and hydrolysis to 3-carbamoyl derivative, reaction with base and then cyanide to form the corresponding 4-cyano derivative followed by hydrolysis.

2.2.15 Epoxidation of itaconate, reaction with cyanide, and hydrolysis.

2.2.16 Condensation of isobutylene with formaldehyde to form 3-methylene-1,5-pentanediol followed by oxidation with nitrogen dioxide and nitric acid.

2.2.17 Oxidation of 1-hydroxy-3-cyclopentenecarboxylic acid derivatives.

2.2.18 Hydration of *cis*- and *trans*-aconitic acids and rearrangement of isocitric and alloisocitric acids by heating in alkali.

3. Physical Properties

3.1 Ionization Constants

Citric acid is a relatively strong organic acid, as indicated by the first dissociation constant of 6.92×10^{-4} at 15°C. Second and third dissociation constants are 1.65×10^{-5} and 4.13×10^{-7}, respectively.

The pKa values at 25°C are pK_1 3.128, pK_2 4.762, and pK_3 6.396. The dissociation constants of citric acid in water at various temperatures are given in Table 1 [2,4-6,41-43].

Table 1

Ionization Constants and pKa Values of Citric Acid in Water at Various Temperatures

Temperature	Ionization constant		
(°C)	$K_1 \times 10^4$ (pKa)	$K_2 \times 10^5$ (pKa)	$K_3 \times 10^7$ (pKa)
0	6.05 (3.218)	1.45 (4.839)	4.05 (6.393)
5	6.31 (3.200)	1.54 (4.812)	4.11 (6.3 86)
10	6.69 (3.175)	1.60 (4.796)	4.14 (6.383)
15	6.92 (3.160)	1.65 (4.783)	4.13 (6.384)
20	7.21 (3.142)	1.70 (4.770)	4.09 (6.388)
25	7.45 (3.128)	1.73 (4.762)	4.02 (6.396)
30	7.66 (3.116)	1.76 (4.754)	3.99 (6.399)
35	7.78 (3.109)	1.77 (4.752)	3.78 (6.423)
40	7.96 (3.099)	1.78 (4.750)	3.69 (6.433)
45	7.99 (3.097)	1.76 (4.754)	3.45 (6.462)
50	8.04 (3.095)	1.75 (4.757)	3.28 (6.484)

3.2 Metal Ion Binding Characteristics

Citric acid is a well-known chelating agent for a wide variety of metal ions [80]. A summary of reporting binding constants is given in Table 3.

3.3 Solubility Characteristics

Citric acid is readily soluble in water, and the solubility of the anhydrous substance at various temperatures in water is shown in Table 3. It is moderately soluble in alcohol, and its solubility in various concentration of aqueous solution of ethyl alcohol is given in Table 4. The solubility at 18°C (g/100 g solvent) in methanol and propyl alcohol is 197 and 62.8, respectively. The solubility in various organic solvents is summarized in Table 5, which shows that citric acid is only sparingly soluble in diethyl ether. The anhydrous acid is insoluble in chloroform, benzene, carbon disulfide, carbon tetrachloride, and toluene [1,2,4-12,41,44].

The heat of solution of citric acid monohydrate is –16.3 kJ/mol (–3.9 kcal/mol) at 25°C [4].

The surface tension of a 0.167 molal solution in contact with air is 69.51 mN/m (dyne/cm) at 30°C. The equivalent conductivity is 8.0×10^{-4} S/cm^2 at 25°C [6].

An aqueous solution of 1% w/v citric acid monohydrate (equivalent to 0.1 N solution) has a pH of 2.2 [2,4]. A practical manual on the formulation and dispensing of pharmaceutical products discloses that the pH of a 1% solution of anhydrous citric acid is about 2.6 [41]. A lemon juice, which contains about 7 percent citric acid, has a pH that is less than 3 [5].

The specific gravities of anhydrous and hydrated citric acid are 1.665 g/cm^3 and 1.542 g/cm^3, respectively [2,4,44]. An account describes the density of anhydrous citric acid as 1.665 g/cm^3 at its melting point of 153°C, and the density of 1.542 g/cm^3 for the hydrated acid after it becomes anhydrous and melts sharply at 153°C [6]. The densities of aqueous solutions of citric acid monohydrate at 15°C are given in Table 6 [2,5,6,43].

Table 2

Binding Constants, Measured at 25°C, of Citric acid with Various Monovalent and Divalent Metal Ions

Metal Ion	log K_1	log K_2
Li(I)	0.83 [a]	–
Na(I)	0.70 [a]	–
K(I)	0.59 [a]	–
Mg(II)	3.25 [b]	0.84 [b]
Ca(II)	3.18 [b]	1.05 [b]
Ba(II)	2.55 [b]	0.79 [b]
Mn(II)	4.15 [a]	2.16 [a]
Fe(II)	4.40 [a]	2.65 [a]
Co(II)	5.00 [a]	3.02 [a]
Ni(II)	5.40 [a]	3.30 [a]
Cu(II)	5.90 [a]	3.42 [a]
Zn(II)	4.98 [a]	2.98 [a]

(a) ionic strength = 0.1

(b) ionic strength = 1.0

Table 3

Solubility of Anhydrous Citric acid in Water

Solid phase	Temperature (°C)	Concentration (wt %)
$C_6H_8O_7.H_2O$	10	54.0
$C_6H_8O_7.H_2O$	20	59.2
$C_6H_8O_7.H_2O$	30	64.3
$C_6H_8O_7.H_2O + C_6H_8O_7$	36.6 [a]	67.3
$C_6H_8O_7$	40	68.6
$C_6H_8O_7$	50	70.9
$C_6H_8O_7$	60	73.5
$C_6H_8O_7$	70	76.2
$C_6H_8O_7$	80	78.8
$C_6H_8O_7$	90	81.4
$C_6H_8O_7$	100	84.0

(a) Transition point.

Table 4

Solubility of Anhydrous and Hydrated Citric acid in Aqueous Ethyl Alcohol Solutions at 25°C

Anhydrous		Ethyl alcohol (wt %)	Monohydrate	
d_{25} (saturated solution)	Solubility (g/100 g saturated solution)		Solubility (g/100 g saturated solution)	d_{25} (saturated solution)
1.297	62.3	20	66.0	1.286
1.246	59.0	40	64.3	1.257
-	-	50	63.3	1.237
1.190	54.8	60	62.0	1.216
1.120	48.5	80	58.1	1.163
1.010	38.3	100	49.8	1.068

Table 5

Solubility of Anhydrous and Hydrated Citric acid in Various Organic Solvents at 25°C

Anhydrous		Solvent	Monohydrate	
d_{25} (saturated solution)	Solubility (g/100 g saturated solution)		Solubility (g/100 g saturated solution)	d_{25} (saturated solution)
0.8861	4.22	amyl acctatc	5.980	0.8917
-	-	amyl alcohol	15.430	0.8774
-	-	ethyl acetate	5.276	0.9175
-	-	diethyl ether	2.174	0.7228
0.7160	1.05	diethyl ether (abs)	-	-
-	-	chloroform	0.007	1.4850

Table 6

Density of Aqueous Citric acid Solutions at 15°C

$C_6H_8O_7 \cdot H_2O$ (wt %)	Concentration of solution g/mL	lb/gal	d_{15}
2	20.15	0.1681	1.0074
4	40.60	0.3388	1.0149
6	61.36	0.5121	1.0227
8	82.47	0.6883	1.0309
10	103.9	0.8673	1.0392
12	125.6	1.049	1.0470
14	147.7	1.232	1.0549
16	170.1	1.420	1.0632
18	192.9	1.610	1.0718
20	216.1	1.803	1.0805
22	239.6	1.999	1.0889
24	263.3	2.198	1.0972
26	287.6	2.400	1.1060
28	312.3	2.606	1.1152
30	337.3	2.815	1.1244
32	362.6	3.026	1.1332
34	388.3	3.241	1.1422
36	414.5	3.460	1.1515
38	441.3	3.682	1.1612
40	468.4	3.909	1.1709
42	496.2	4.141	1.1814
44	523.6	4.369	1.1899
46	551.9	4.605	1.1998
48	580.9	4.848	1.2103
50	610.2	5.092	1.2204
52	640.0	5.341	1.2307
54	670.1	5.593	1.2410
56	700.8	5.848	1.2514
58	732.4	6.112	1.2627
60	764.3	6.378	1.2738
62	796.6	6.648	1.2849
64	829.4	6.922	1.2960
66	862.7	7.199	1.3071

Figure 2. Morphology of citric acid anhydrate, obtained using optical microscopy at a magnification of 40x.

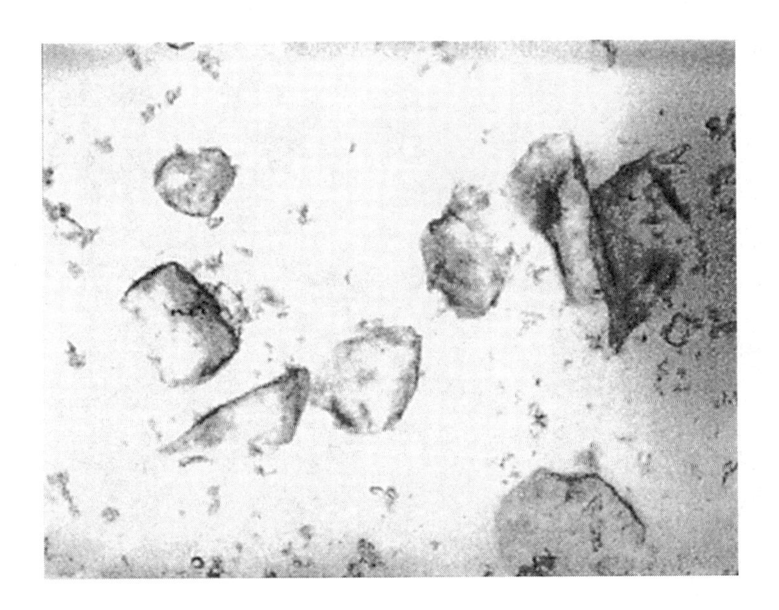

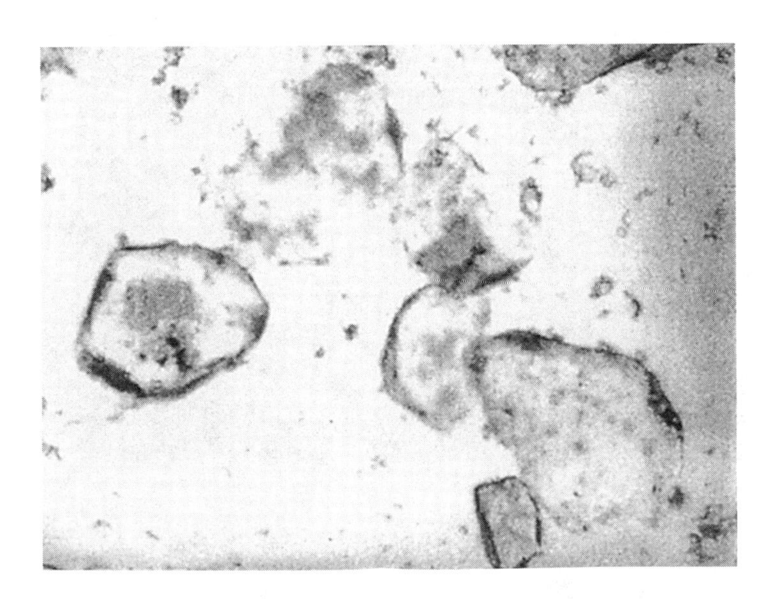

Figure 3. Morphology of citric acid monohydrate, obtained using scanning electron microscopy at (a) 35x and (b) 350x.

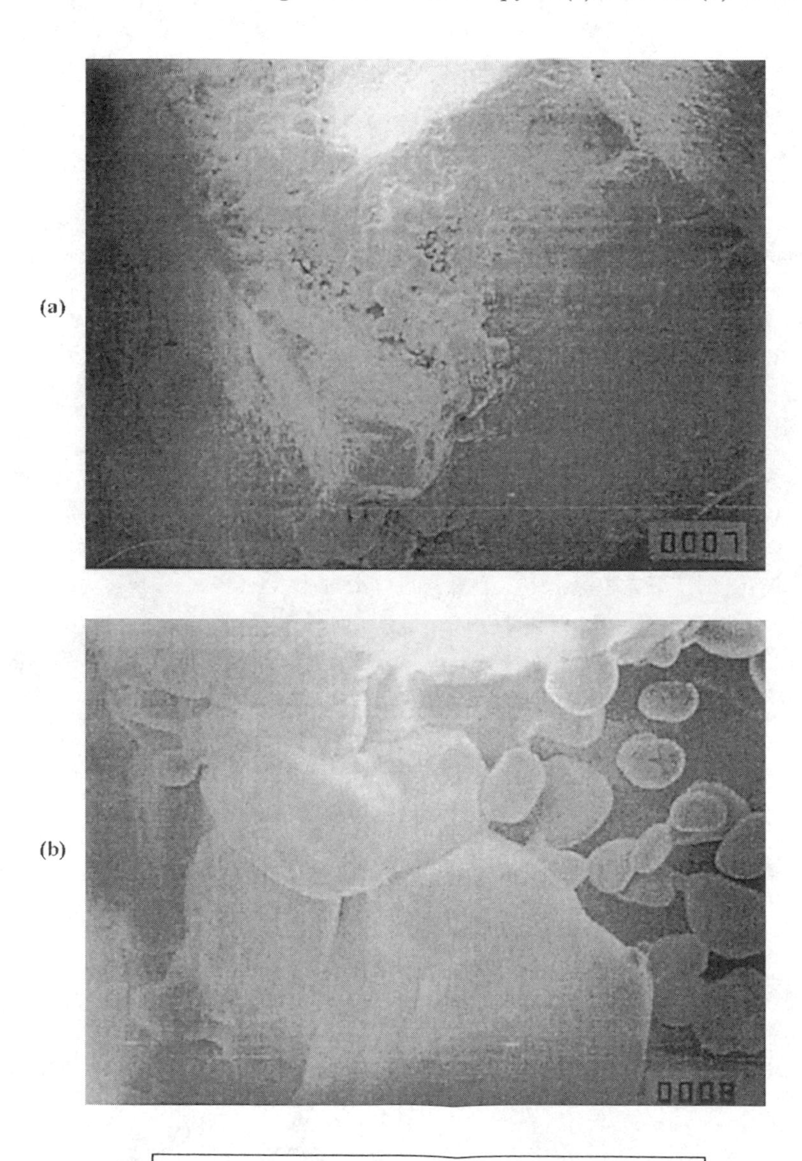

Figure 4. Molecular structure of citric acid monohydrate.

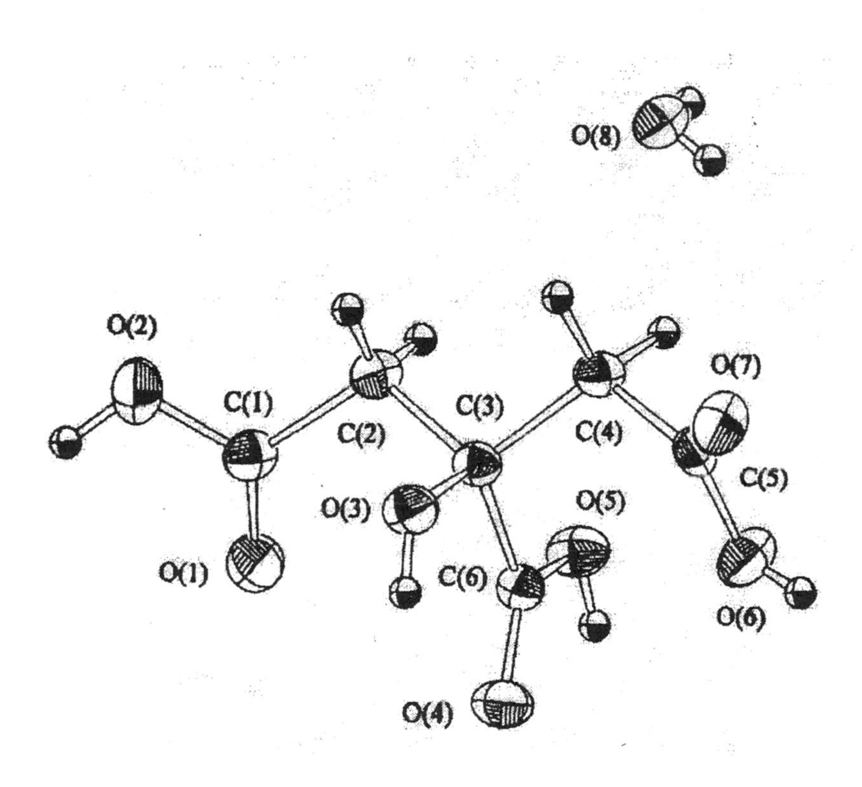

Table 7

Crystallographic Data for Citric Acid Anhydrate

Empirical formula	$C_6H_8O_{10}$
Formula weight	210.14
Crystal color, habit	Colorless, prismatic
Crystal dimensions	0.60 × 0.40 × 0.30 mm
Crystal system	Orthorhombic
Lattice type	Primitive
Lattice parameters	a = 9.3196 (8) Å
	b = 15.400 (2) Å
	c = 6.2963 (9) Å
	V = 903.7 (2) Å^3
Space group	$P\ 2_1\ 2_1\ 2_1$
Z	4
D_c (g cm^{-3})	1.544

Table 8

Positional and Thermal Parameters for Citric Acid

Atom	X	Y	Z	B_{eq} (Å^2)
O(1)	1.2340 (1)	0.35624 (7)	0.3093 (2)	3.87 (2)
O(2)	1.2893 (1)	0.44622 (8)	0.5733 (2)	4.03 (3)
O(3)	0.9345 (1)	0.47234 (6)	0.3556 (2)	2.76 (2)
O(4)	0.9461 (1)	0.34613 (6)	0.0685 (2)	3.33 (2)
O(5)	0.9494 (1)	0.24133 (6)	0.3143 (2)	3.15 (2)
O(6)	0.6539 (1)	0.33335 (8)	0.2727 (2)	3.60 (2)
O(7)	0.5771 (1)	0.44538 (6)	0.4657 (2)	3.32 (2)
O(8)	0.9475 (1)	0.12522 (7)	0.9935 (2)	3.05 (2)
C(1)	1.2040 (1)	0.39183 (9)	0.4726 (2)	2.66 (3)
C(2)	1.0629 (2)	0.37791 (9)	0.5867 (2)	2.57 (3)
C(3)	0.9348 (1)	0.38633 (8)	0.4371 (2)	2.24 (2)
C(4)	0.7954 (2)	0.37198 (9)	0.5655 (2)	2.61 (3)
C(5)	0.6639 (1)	0.38839 (9)	0.4336 (2)	2.57 (3)
C(6)	0.9446 (2)	0.32233 (8)	0.2499 (2)	2.43 (3)
H(1)	1.355 (1)	0.4705 (9)	0.498 (2)	2.340
H(2)	1.050 (1)	0.4266 (8)	0.681 (2)	2.340
H(3)	1.072 (1)	0.3104 (8)	0.647 (2)	2.340
H(4)	0.938 (2)	0.4772 (9)	0.227 (2)	2.340
H(5)	0.805 (1)	0.4156 (8)	0.681 (2)	2.340
H(6)	0.788 (1)	0.2955 (8)	0.636 (2)	2.340
H(7)	0.953 (2)	0.2173 (9)	0.201 (2)	2.340
H(8)	0.590 (2)	0.332 (1)	0.212 (2)	2.340
H(9)	0.981 (1)	0.0519 (8)	1.033 (2)	2.340
H(10)	0.896 (2)	0.116 (1)	0.930 (3)	2.340

Table 9

Bond Distances in Citric Acid

Atomic bond	Bond distance	Atomic bond	Bond distance
O(1)–C(1)	1.198 (6)	O(2)–C(1)	1.317 (6)
O(2)–H(1)	0.86 (5)	O(3)–C(3)	1.421 (5)
O(3)–H(4)	0.82 (5)	O(4)–C(6)	1.200 (6)
O(5)–C(6)	1.312 (5)	O(5)–H(7)	0.80 (5)
O(6)–C(5)	1.324 (6)	O(6)–H(8)	0.71 (5)
O(7)–C(5)	1.211 (6)	O(8)–H(9)	1.20 (5)
O(8)–H(10)	0.64 (6)	C(1)–C(2)	1.514 (7)
C(2)–C(3)	1.526 (7)	C(2)–H(2)	0.96 (5)
C(2)–H(3)	1.11 (5)	C(3)–C(4)	1.546 (7)
C(3)–C(6)	1.539 (7)	C(4)–C(5)	1.502 (7)
C(4)–H(5)	0.99 (5)	C(4)–H(6)	1.26 (5)

Table 10

Bond Angles and Torsion Angles for Citric Acid

Atomic bond	Bond angle	Atomic bond	Bond angle
C(1)–O(2)–H(1)	116 (4)	C(3)–O(3)–H(4)	116 (4)
C(6)–O(5)–H(7)	99 (4)	C(5)–O(6)–H(8)	119 (4)
H(9)–O(8)–H(10)	97 (6)	O(1)–C(1)–O(2)	124.3 (5)
O(1)–C(1)–C(2)	123.0 (5)	O(2)–C(1)–C(2)	112.7 (4)
C(1)–C(2)–C(3)	112.0 (5)	C(1)–C(2)–H(2)	107 (3)
C(1)–C(2)–H(3)	103 (3)	C(3)–C(2)–H(2)	103 (3)
C(3)–C(2)–H(3)	111 (3)	H(2)–C(2)–H(3)	122 (4)
O(3)–C(3)–C(2)	107.7 (4)	O(3)–C(3)–C(4)	108.7 (4)
O(3)–C(3)–C(6)	108.7 (4)	C(2)–C(3)–C(4)	108.8 (4)
C(2)–C(3)–C(6)	111.9 (4)	C(4)–C(3)–C(6)	111.0 (4)
C(3)–C(4)–C(5)	111.9 (4)	C(3)–C(4)–H(5)	102 (3)
C(3)–C(4)–H(6)	111 (2)	C(5)–C(4)–H(5)	111 (3)
C(5)–C(4)–H(6)	108 (2)	H(5)–C(4)–H(6)	112 (3)
O(6)–C(5)–O(7)	123.0 (5)	O(6)–C(5)–C(4)	111.9 (4)
O(7)–C(5)–C(4)	125.1 (5)	O(4)–C(6)–O(5)	125.7 (5)
O(4)–C(6)–C(3)	122.3 (4)	O(5)–C(6)–C(3)	112.0 (4)

Atomic bond	Torsion angle	Atomic bond	Torsion angle
O(1)–C(1)–C(2)–C(3)	–48.7 (6)	O(2)–C(1)–C(2)–C(3)	131.8 (5)
O(3)–C(3)–C(2)–C(1)	–62.0 (5)	O(3)–C(3)–C(4)–C(5)	57.1 (5)
O(3)–C(3)–C(6)–O(4)	–2.3 (7)	O(3)–C(3)–C(6)–O(5)	178.2 (4)
O(4)–C(6)–C(3)–C(2)	–121.1 (6)	O(4)–C(6)–C(3)–C(4)	117.2 (6)
O(5)–C(6)–C(3)–C(2)	59.4 (6)	O(5)–C(6)–C(3)–C(4)	–62.3 (6)
O(6)–C(5)–C(4)–C(3)	63.2 (5)	O(7)–C(5)–C(4)–C(3)	–115.9 (5)
C(1)–C(2)–C(3)–C(4)	–179.6 (4)	C(1)–C(2)–C(3)–C(6)	57.4 (5)
C(2)–C(3)–C(4)–C(5)	174.1 (4)	C(5)–C(4)–C(3)–C(6)	–62.4 (5)

3.6.2 X-Ray Powder Diffraction Pattern

The x-ray powder diffraction pattern of citric acid USP (anhydrate phase) was obtained using a Philips P7100 system. The powder pattern is shown in Figure 5, while a summary of the observed scattering angles, d-spacings, and relative intensities is found in Table 11.

X-ray diffractometer has been used to determine the composition in bulk-prepared solidified citric acid melt, using a powdered sample as the reference standard. The powder pattern of anhydrous citric acid and bulk-prepared solidified citric acid melt indicated that the bulk-prepared citric acid melt consisted of an amorphous and crystalline citric acid mixture [40].

3.7 Hygroscopicity

At a temperature of 25°C, anhydrous citric acid absorbs insignificant amounts of water when stored at relative humidities between about 25% and 50%. However, at relative humidities in the range of 50 – 75%, it absorbs significant amounts and the monohydrate phase is formed at relative humidities approaching 75%. At relative humidities greater than 75% substantial amounts of water are absorbed by the monohydrate.

The hydrated acid is slightly deliquescent in moist air. At relative humidities less than about 65%, citric acid monohydrate effloresces at 25°C. The anhydrous acid is formed at relative humidities less than about 40%. At relative humidities in the range of 65 – 75%, citric acid monohydrate absorbs insignificant amounts of moisture, but under more humid conditions substantial amounts of water are absorbed [1,2,4,6-11].

3.8 Thermal Methods of analysis

3.8.1 Melting Behavior

The melting point of the anhydrous form is 153°C. Citric acid monohydrate loses its water of crystallization in dry air, in a vacuum over sulfuric acid, or when heated at about 40 to 50°C. On gentle heating, the monohydrate crystals soften at about 75°C with the loss of water, and finally melt completely in the range of 135 – 150°C. On rapid heating, the crystals melt at 100°C, solidify as they become anhydrous, and melt sharply at 153°C [2,4-6,11,12,44,45].

Figure 5. X-ray powder diffraction pattern of citric acid USP (anhydrate phase).

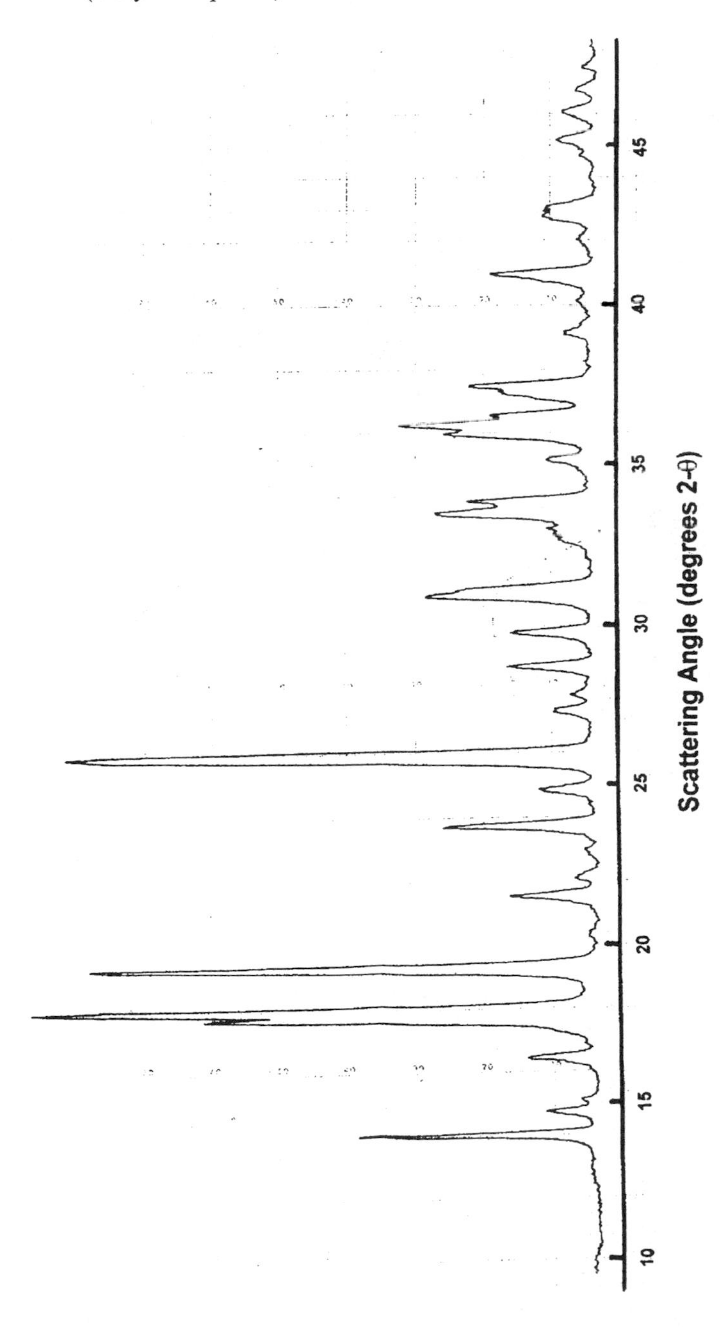

Table 11

Crystallographic Parameters Derived from the X-Ray Powder Pattern of Citric Acid Anhydrate

ID	Scattering Angle (degrees 2-θ)	d-spacing (Å)	Relative Intensity (%)
1	14.0556	6.3100	42.00
2	14.7778	6.0032	9.33
3	15.1444	5.8586	3.33
4	16.4444	5.3983	12.00
5	17.7000	5.0181	68.00
6	17.9444	4.9503	100.00
7	19.2778	4.6108	88.00
8	20.3333	4.3738	1.33
9	21.5556	4.1285	15.33
10	22.1111	4.0260	4.00
11	23.0000	3.8724	2.00
12	23.7222	3.7561	26.67
13	24.8889	3.5826	10.00
14	25.9222	3.4421	92.00
15	27.3333	3.2675	7.33
16	27.8333	3.2100	4.33
17	28.7222	3.1126	15.33
18	29.7778	3.0046	15.33
19	30.9444	2.8940	29.33
20	31.1111	2.8788	24.00
21	32.5556	2.7543	6.00

Table 11 (continued)

Crystallographic Parameters Derived from the X-Ray Powder Pattern of Citric Acid Anhydrate

ID	Scattering Angle (degrees 2-θ)	d-spacing (Å)	Relative Intensity (%)
22	33.0000	2.7183	8.00
23	33.5000	2.6788	28.00
24	33.8889	2.6490	22.67
25	35.1667	2.5556	8.33
26	36.0000	2.4983	26.33
27	36.2778	2.4798	34.00
28	36.5556	2.4616	18.00
29	37.3333	2.4121	16.67
30	37.5000	2.4018	22.00
31	39.1667	2.3033	5.33
32	40.1667	2.2483	2.67
33	41.0000	2.2045	18.00
34	42.0000	2.1543	3.00
35	42.7778	2.1169	9.00
36	43.0000	2.1065	8.00
37	44.0000	2.0609	0.67
38	44.7222	2.0293	2.67
39	45.1889	2.0094	6.67
40	46.1111	1.9714	5.33
41	46.8111	1.9435	3.33
42	47.5000	1.9169	2.67

3.8.2 Differential Scanning Calorimetry

Differential scanning calorimetry was obtained on citric acid USP (anhydrate phase) using a TA Instruments model 9020. The thermogram is shown in Figure 6, and consists of a single endotherm identified as the melting phase transition. The onset of melting was noted at 154.9°C, and the peak maximum was observed at 157.2°C. The enthalpy of fusion computed for this sample was calculated as 229 J/g.

3.8.3 Heat of Combustion

The heat of combustion of anhydrous citric acid and its hydrated form are –1985 kJ/mol (–474.5 kcal/mol) and –1972 kJ/mol (–471.4 kcal/mol), respectively [4]. Another account reports the heats of combustion at 25°C, for finding 1.96 MJ/mol (468.5 kal/mol) and 1.952 MJ/mol (466.6 kal/mol) for citric acid anhydrous and monohydrate, respectively [6].

3.9 Spectroscopy

3.9.1 UV/VIS Spectroscopy

The ultraviolet absorption spectra of citric acid in methanol and in 0.001 N HCl (each at concentration of 960 ppm) were recorded using a Hitachi U-3210 spectrophotometer, and are shown in Figure 7. The spectra were not found to be greatly affected by the nature of the solvent used. An absorbance maximum of 207.0 nm was noted in the 0.001 N HCl solvent, while a maximum at 210.2 nm was found for the methanolic solution. The absorbencies within these two solutions were found to be 1.1645 and 0.9705, respectively. Therefore, the respective molar absorptivities are computed to be 240 and 200 liter/cm·mole, which are classified as being low-intensity absorptions.

3.9.2 Vibrational Spectroscopy

The infrared absorption spectrum of citric acid was obtained in a potassium bromide pellet (using approximately 2 mg of citric acid dispersed in 200 mg KBr), and recorded on a JASCO FTIR-200 spectrophotometer. The spectrum thusly obtained is shown in Figure 8, and the assignment of the characteristic bands is given in Table 12.

Figure 6. Differential scanning calorimetry thermogram of citric acid USP (anhydrate phase).

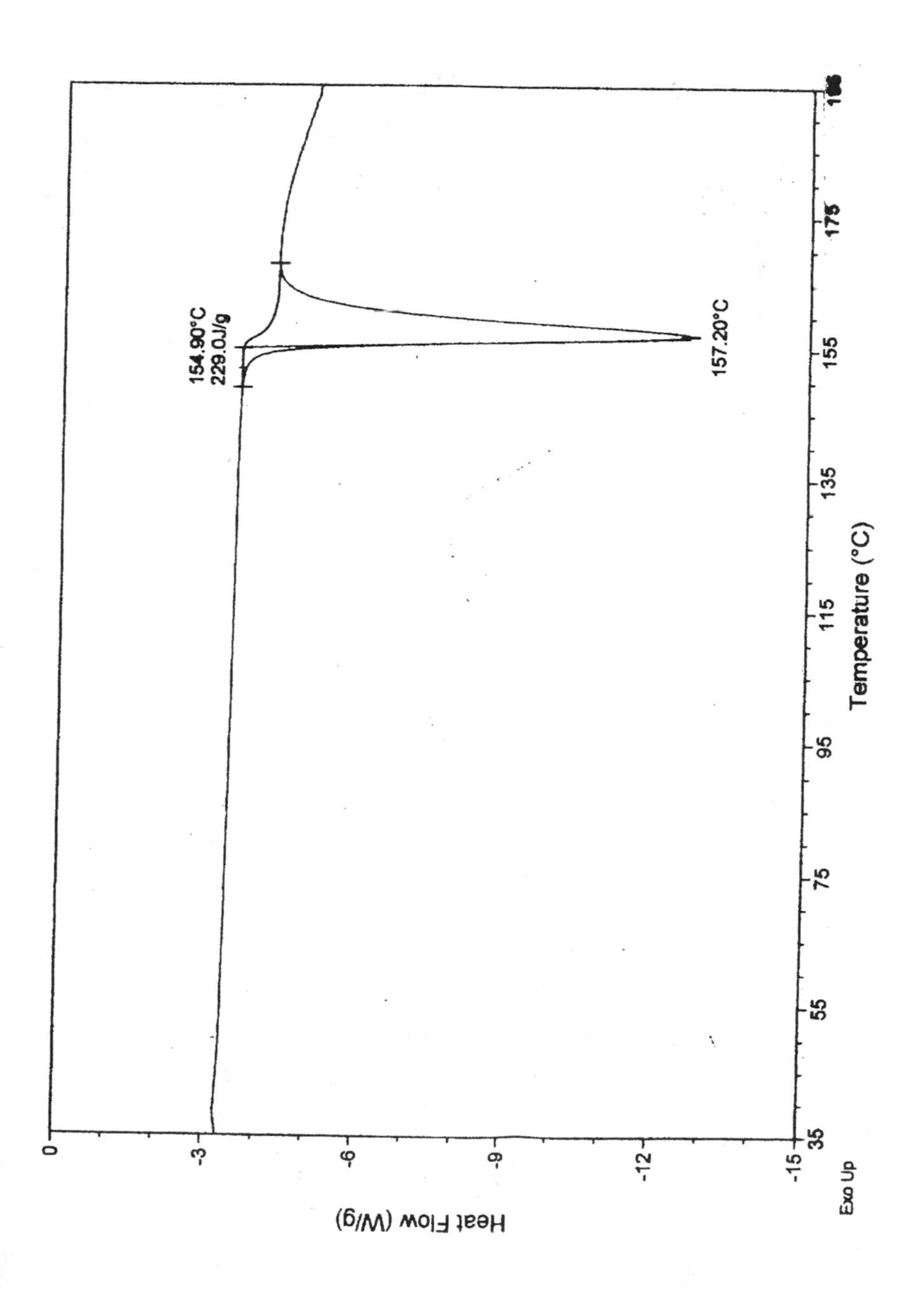

Figure 7. Ultraviolet absorption spectrum of citric acid in methanol and in 0.001 N HCl.

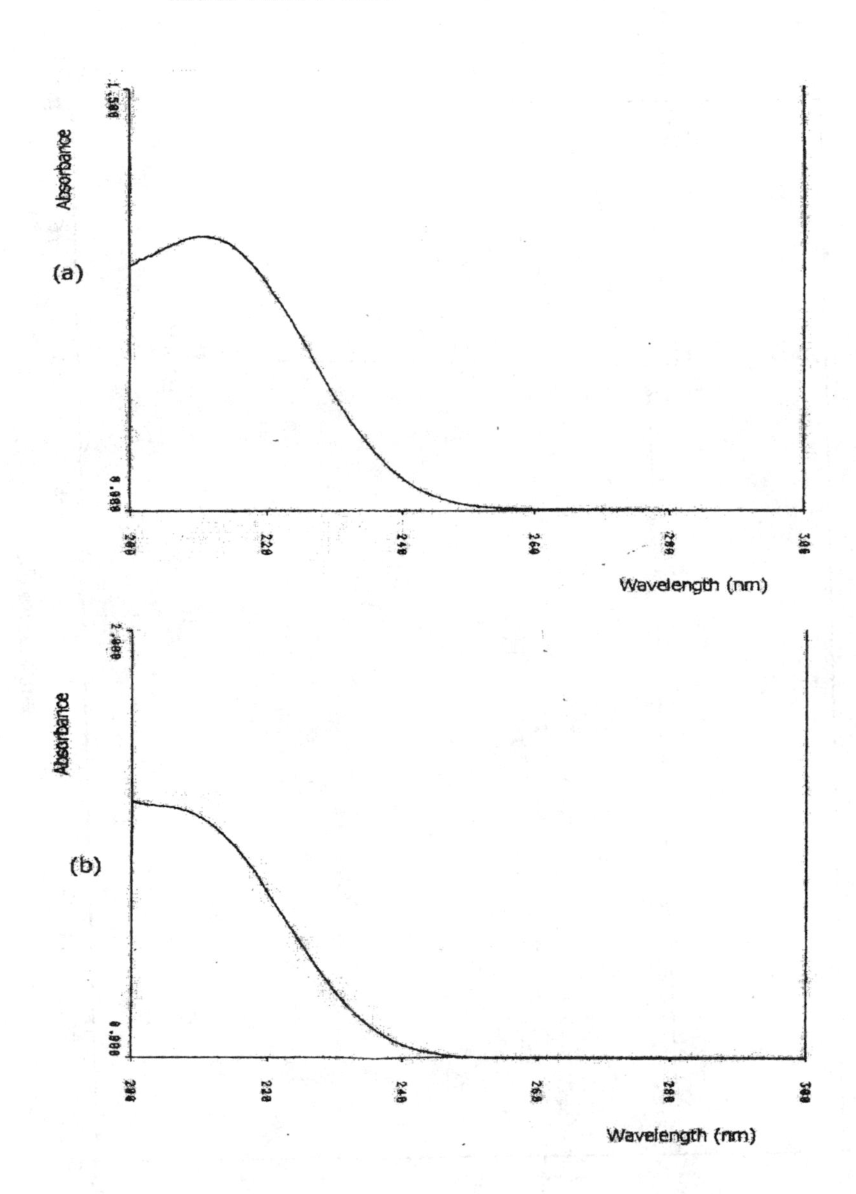

Figure 8. Infrared absorption spectrum of citric acid.

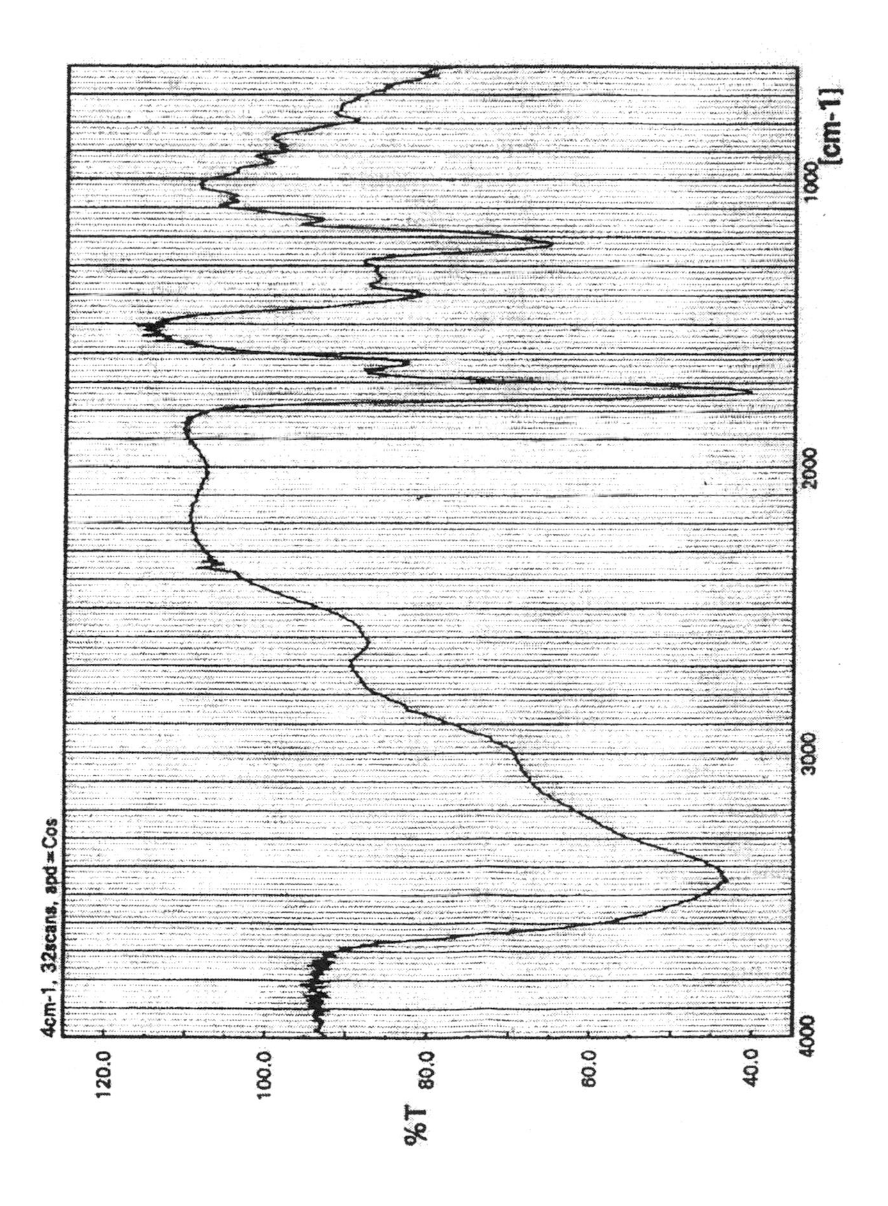

Table 12

Assignment for the Characteristic Infrared Absorption Bands of Citric acid

Energy (cm^{-1})	Band Assignment
3600–2900	Carboxyl –OH stretching mode
3450	Hydroxyl –OH stretching mode
2900–3000	–C–H stretching mode
1730	–C=O stretching mode
1620	–C–H bending mode
1400	C–O–H in-plane bending mode
1220	–C–O stretching mode

3.9.3 Nuclear Magnetic Resonance Spectrometry

3.9.3.1 ^{1}H-NMR Spectrum

The 270 MHz ^{1}H-NMR spectrum of citric acid was obtained in deuterated methanol using a JEOL JNM LA-270 spectrometer, and is shown in Figure 9. All chemical shifts are reported as δ values (ppm) relative to tetramethylsilane (TMS) and residual CD_3OD as internal standards. The assignments for the observed bands are found in Table 13, together with multiplicities and coupling constants.

3.9.3.2 ^{13}C-NMR Spectrum

The broadband decoupled ^{13}C-NMR spectrum of citric acid is shown in Figure 10. The spectrum was recorded at 24°C in methanol-d using a JEOL JNM LA-270 spectrometer operating at 67.8 MHz. Chemical shifts were measured relative to TMS, and assignments for the observed peaks are found in Table 14.

3.10 Mass Spectrometry

The mass spectrum of citric acid was recorded on a JEOL JMS D-600 spectrometer, using the chemical ionization (CI) with isobutane and electron impact (EI) methods. The obtained CI and EI mass spectra are shown in Figures 11 and 12, respectively. Assignments for the main observed fragments are presented in Table 15.

4. Methods of Analysis

4.1 Compendial Tests

Citric Acid USP, as the anhydrate or monohydrate phase, is specified to contain not less than 99.5 percent and not more than 100.5 percent of the substance, calculated on the anhydrous basis. It is to be preserved in tight containers. Labeling is suggested to indicate whether it is anhydrous or hydrous. A number of compendial tests have been established by the USP 24 and other pharmacopoeial compendia.

Figure 9. ¹H-NMR spectrum of citric acid.

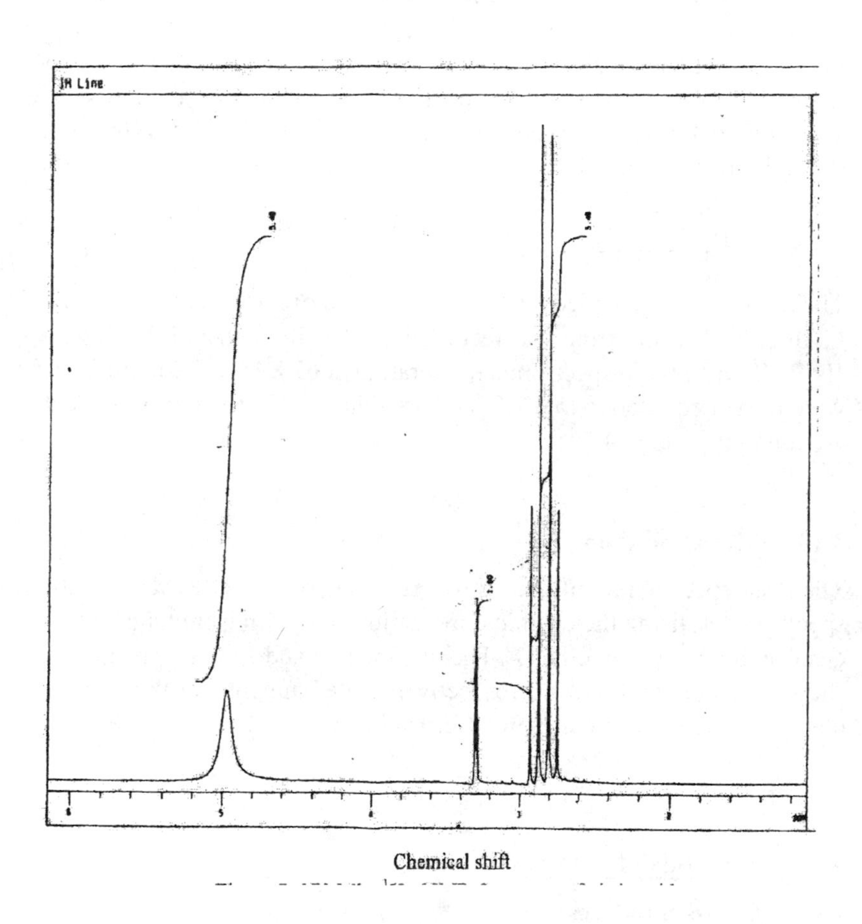

Figure 10. ^{13}C-NMR spectrum of citric acid.

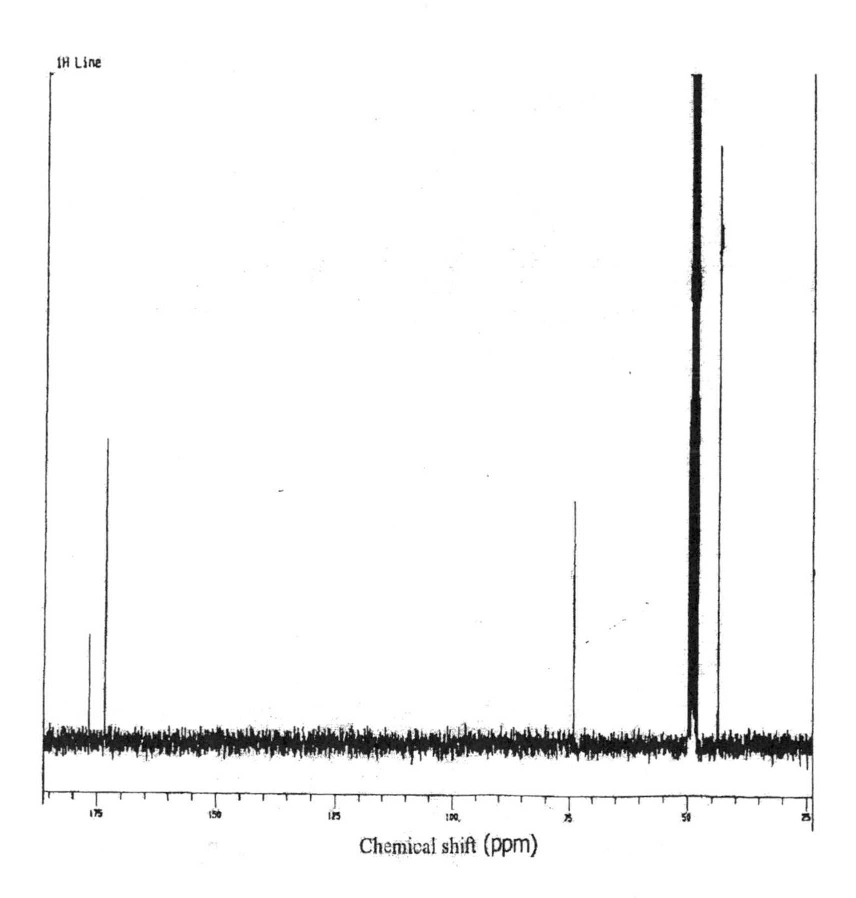

Table 13

Assignments for the Resonance Bands in the ^{1}H-NMR Spectrum of Citric acid

Chemical Shift (ppm)	Multiplicity	Number of Proton	Assignment
4.97	Singlet (broad)	4	-COO***H***, -O***H***
2.84	Double-doublet (J=15.7, 33.5 Hz)	4	-C***H***$_2$C(OH)(COOH)C***H***$_2$-

Table 14

Assignments for the Resonance Bands in the ^{13}C-NMR Spectrum of Citric acid

Chemical Shift (ppm)	Carbon Number
176.77	C(6)
173.47	C(1), C(5)
74.14	C(3)
43.83	C(2), C(4)

Figure 11. Chemical ionization mass spectrum of citric acid.

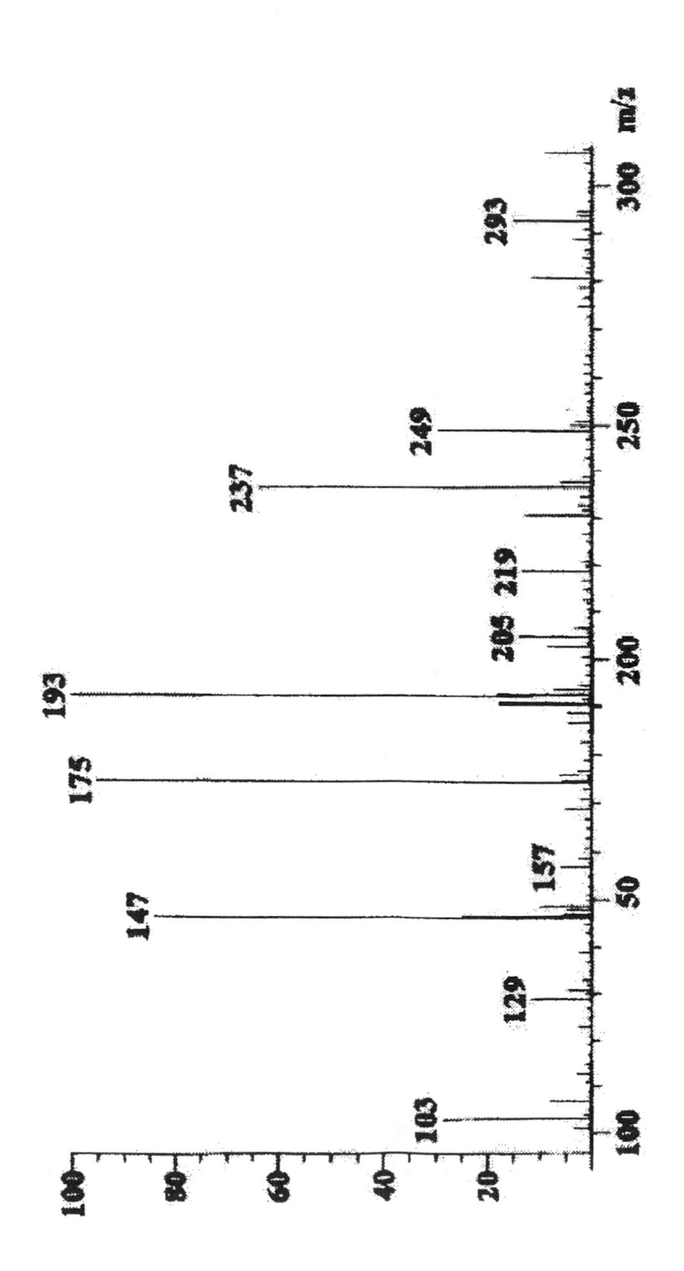

Figure 12. Electron impact mass spectrum of citric acid.

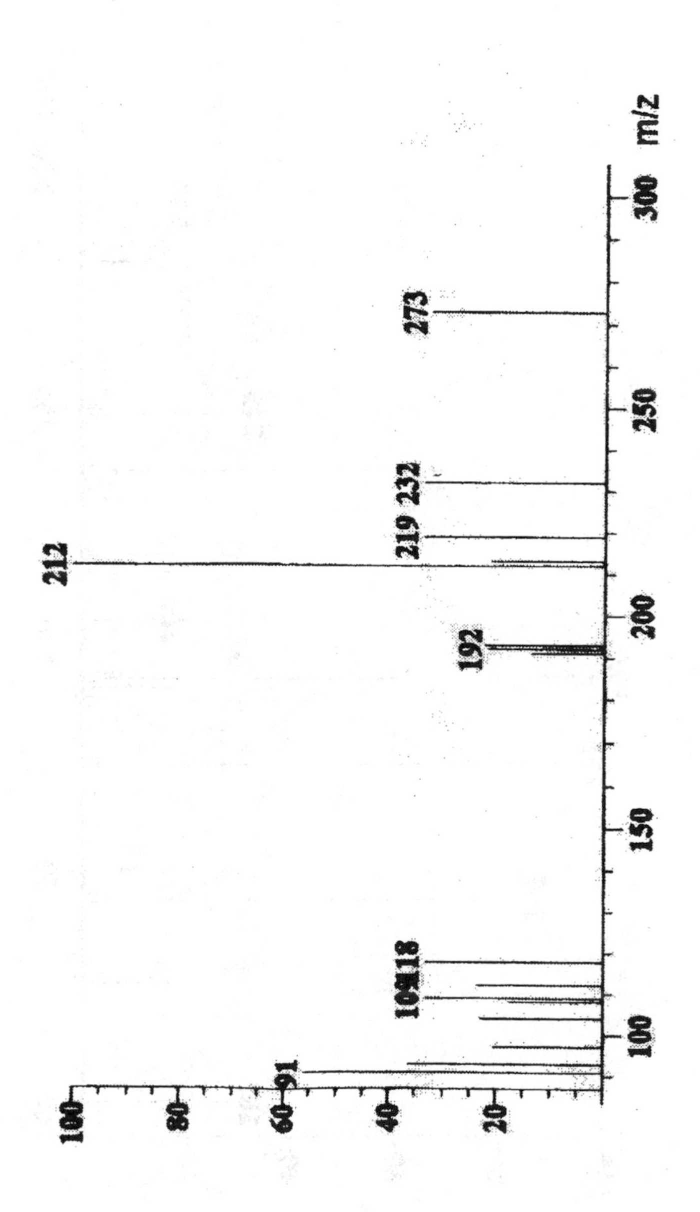

Table 15

Electron Impact (EI) and Chemical Ionization (CI) Mass Spectral Data for Citric acid

MS Mode	m/z Ratio (% relative intensity)	Fragment Assignment
EI	212 (100)	$[M + (18) H_2O + (2) H_2]^+$
	192 (22.7)	M^+
	118 (34.1)	$[M - C(OH)COOH]^+$
CI (Isobutane)	249 (29.5)	$[M + 57 (C_4H_9)]^+$
	237 (63.6)	$[M + 1 (H) + 42 (C_3H_6) + 2 (H_2)]^+$
	193 (100)	$[M + 1]^+$
	175 (95.4)	$[M + 1 (H) - 18 (H_2O)]^+$
	147 (84.1)	$[M + 1 (H) - 44 (CO_2) - 2 (H_2)]^+$
	103 (29.5)	$[M + 1 (H) - 90 (2 \times CO_2H)]^+$

4.1.1 Identification

According to the United States Pharmacopoeia 24 [46], citric acid is identified by using a solution of the analyte responds to the test for *Citrate* (pyridine - acetic anhydride test) <191>. In addition to the citrate test, the Chinese pharmacopoeia [8] includes the test of igniting the crystal that should decompose gradually, but for which no caramel odor is perceptible (this is a distinction from tartaric acid).

Identification test in to the British Pharmacopoeia [7] are arranged in two tiers, classified as the: first identification (tests B and E) and the second identification (tests A, C, D, and E).

Test A: About 1 g of the acid sample is dissolved in 10 mL of water. The resulting solution is strongly acidic.

Test B: Examination of the acid sample by infrared absorption spectrophotometry. Compare the spectrum of the analyte with the spectrum obtained with citric acid reference standard after drying both substances at 100°C to 105°C for 24 hours.

Test C: About 5 mg of the acid sample is added to a mixture consisting of 1 mL of acetic anhydride and 3 mL of pyridine. A red color develops.

Test D: About 0.5 g of the acid sample is dissolved in 5 mL of water. The resulting solution is neutralized using 1M sodium hydroxide (about 7 mL), 10 mL of calcium chloride is added, and the solution heated to boiling. A white precipitate is formed.

Test E: The sample under examination complies with the test for water.

4.1.2 Water Content

The water content of citric acid is tested according to USP [46] general test <921>, Method I. The permitted water content is not more than 0.5% for the anhydrate, and not more than 8.8% for the hydrate form.

4.1.3 Readily Carbonizable Substances

Any carbonizable material present in a sample of citric acid is examined by the following procedure [46]. A 1.0 g sample of powdered acid is transferred to a 22 × 175-mm test tube, previously rinsed with 10 mL of sulfuric acid TS and allowed to drain for 10 minutes. 10 mL of sulfuric acid TS is added, and the mixture is agitated until dissolution is complete. The test tube is immersed in a water bath at 90 ± 1°C for 60 ± 0.5 minutes, and the level of the acid solution is kept below the level of the water during the entire period. Then the tube is cooled in running water, and the acid solution is transferred to a color comparison tube. The color of the acid is not darker than that of a similar volume of Matching Fluid K, which contains one part of cobaltous chloride colorimetric solution (C.S.) to nine parts ferric chloride (C.S.). Following the directions of general test <631>, the tubes are observed vertically against a white background.

4.1.4 Residue on Ignition

When examined according to USP general procedure <281>, the ash obtained for an acid sample may exceed 0.05% [46].

4.1.5 Sulfate

The sulfate content of a citric acid sample is determined by adding 1 mL of barium chloride TS, to which has been added 1 drop of hydrochloric acid, to a 10 mL solution (1 in 100) of the analyte. No turbidity owing to precipitation of barium sulfate is produced [46].

4.1.6 Arsenic

The arsenic content of citric acid is determined using USP general procedure <211>, Method I [46]. The maximum level of arsenic permissible is 3 ppm.

4.1.7 Heavy Metals

The heavy metal content of citric acid is determined using USP general procedure <231> [46]. The limit on heavy metals in the substance is 0.001%.

4.1.8 Limit of Oxalate

The oxalate content of citric acid is determined using the following procedure [46]. To 10 mL of a 1 in 10 solution of the acid, which has been neutralized with 6 N ammonium hydroxide, add 5 drops of 3 N hydrochloric acid. The mixture is cooled, and then 2 mL of calcium chloride TS is added. No turbidity owing to precipitation of calcium oxalate is produced.

4.1.9 Organic Volatile Impurities

When examined according to USP general procedure <467>, Method IV, the organic volatile content of a citric acid sample must meet the requirements [46].

4.1.10 Bacterial Endotoxins

If the citric acid under consideration is intended for use in the manufacture of parenteral dosage forms without an appropriate procedure for the removal of bacterial endotoxins, the acid sample may contain not more than 0.5 I.U of endotoxin per milligram, as regulated by the British Pharmacopoeia [7].

4.1.11 Aluminum

According to the British Pharmacopoeia [7], if the citric acid under consideration is intended for use in the manufacture of dialysis solutions, the acid sample should comply with the test for aluminum. About 20 g of the analyte is dissolved in 100 mL of water, and 10 mL of acetate buffer (pH 6.0) is added. The solution should comply with the limit test for aluminum (NMT 0.2 ppm). A mixture of 2 mL of aluminum standard solution (2 ppm Al), 10 mL of acetate buffer solution pH 6.0, and 98 mL of water is used as the reference solution. A mixture of 10 mL of acetate buffer solution pH 6.0 is used to prepare the blank.

4.1.12 Assay

Citric acid is assayed by titration with 1 N sodium hydroxide VS to a pink phenolphthalein endpoint. The procedure calls for the dissolution of about 3 g of accurately weighed sample in 40 mL of water. Each milliliter of 1N NaOH is equivalent to 64.04 mg of citric acid.

4.2 Titrimetric Analysis

Citric acid can be titrated directly with 1 N alkali using thymol blue solution as the indicator [47]. This yields a more distinct end-point, and its use is preferable to that of phenolphthalein. Each milliliter of 1 N alkali is equivalent to 64.04 mg of anhydrous acid, or to 70.05 mg of the monohydrate.

4.2.1 Methods of the United States and Indonesian Pharmacopoeias

Both the Indonesian Pharmacopoeia 1995 [10] and the Unites States Pharmacopoeia 24 [46] use the same procedure. About 3 g of citric acid (accurately weighed) is dissolved in 40 mL of water, phenolphthalein TS is added, and the resulting solution is titrated with 1 N sodium hydroxide to a pink end-point.

4.2.2 Method of the Pharmacopoeia of the Peoples' Republic of China

In this procedure established by the Pharmacopoeia of the Peoples' Republic of China 1997 [8], about 1.5 g of citric acid sample is accurately weighed and dissolved in 40 mL of freshly boiled and cooled water. Three drops of phenolphthalein is added to this solution, and this is then titrated with 1 mol/L sodium hydroxide until a pink color is obtained.

4.2.3 Methods of the British and Indian Pharmacopoeias

According to the British [7] and Indian [9] Pharmacopoeias, about 550 mg of analyte is accurately weighed and dissolved in 50 mL of water, or an accurately weighed of 2 g of analyte is dissolved in 100 mL of water. Each solution is titrated with 1 M sodium hydroxide using phenolphthalein solution as the indicator. Each milliliter of 1 M sodium hydroxide is equivalent to 64.03 mg of citric acid.

4.2.4 Potentiometric Titration

In this method, the position of the endpoint is determined to a potentiometric endpoint, and an indicator substance is not used. The selected indicator sensor should determine the consumption of reagent at the point where the titration curve exhibits a sharp change in the measuring cell voltage (the potential break). The end of the titration corresponds to this amount, which can be found graphically or numerically. One potentiometric titration of citric acid is based on the ability of this organic substance to form a complex with Cu(II), with the enpoint being determined by the appropriate ion selective electrode [48].

Another procedure uses pyrolysis of the organic substance in a stream of ammonia, which results in conversion of carbon into hydrogen cyanide, and that is absorbed in methanolic KOH. The cyanide ion content is found by potentiometric titration with an $AgNO_3$ solution, and an ion-selective electrode based on Ag_2S as the indicator electrode [49].

4.3 Electrochemical Analysis

The direct potentiometric measurement includes procedures in which the measured potential for an electrochemical cell yields the concentration (or activity) of a given ion in solution, or its logarithm. Two distinct techniques are used in which the concentration of the analyte is measured continuously in a flowing liquid or gas. In the first, the sample is drawn into the flowing system and is segmented by air bubbles (Continuous Flow Analysis, CFA). In the second, the sample is injected into the carrier system through a sampling valve (Flow Injection Analysis, FIA) [48].

The method of elemental analysis uses ion selective electrodes (ISE) for the determination of elements in organic substances. The usual decomposition procedures are used, and measurement with the ISE is used as the final step to simplify the determination of the decomposition products. In the determination of carbon, the citric acid is mineralized in a stream of oxygen and volatile decomposition products are passed over a suitable catalyst on which the carbon is converted quantitatively into carbon dioxide. This is absorbed by sodium hydroxide, the absorption solution is acidified with hydrochloric acid, and the partial pressure of the released CO_2 is measured potentiometrically with a suitable gas electrode [50].

A modified procedure applying a gas electrode for the determination of CO_2 can also be used after mineralization of citric acid by the wet method [51]. A different procedure allowing the CO_2 formed to be absorbed as $BaCO_3$ in an alkaline $BaCl_2$ solution is also reported [52].

Since an ISE mostly cannot be used for the direct determination of functional groups in organic substances (unless these substances are first converted into ionic species), direct potentiometry can be used only exceptionally. Very simple sensors can be used in the FIA technique, with citric acid forming complexes with Cu(II) that can be determined with acceptable precision by using an indicator sensor consisting of a copper wire [53].

A bio-electrochemical determination of citric acid in fruits, juices, and sport drinks was reported by Prodromidis *et al* [54]. The method is based on the action of immobilized citrate lyase, oxaloacetate decarboxylase, and pyruvate oxidase enzymes, which convert citric acid into hydrogen peroxide. The H_2O_2 is monitored amperometrically with a H_2O_2 probe. A multi-membrane system, consisting of a cellulose acetate membrane for elimination of interferences, an enzyme membrane, and a protective polycarbonate membrane were placed on a Pt electrode and used with a fully automated flow injection manifold. Average recoveries ranged from 96% to 104%.

4.4 Spectroscopic Analysis

4.4.1 Colorimetry

The 16th edition of the AOAC described a determination of citric acid in cheese [55]. The procedure begins with dispersion of the sample in water, with the protein being precipitated by trichloroacetic acid. After filtration of the precipitate, the filtrate is treated with pyridine and acetic anhydride according to the widely used Furth and Herrmann reaction [6]. The resulting yellow solution of the citric acid complex is measured at 428 nm.

4.4.2 Spectrophotometry

A handbook of food analysis suggested that citric acid is best determined enzymatically [56]. Following the AOAC enzymatic procedure [55], citric acid is converted by citric lyase to oxaloacetate and acetate. In the presence of the malate dehydrogenase and lactate dehydrogenase enzymes, oxaloacetate and its decarboxylation product pyruvate are reduced to L-malate and L-lactate, respectively, by reduced nicotinamide-adenine dinucleotide (NADH). The amount of NADH oxidized is stoichiometric with the amount of citrate present. The NADH consumed is measured at 340 nm with a spectrophotometer, or 365 nm with Hg filter photometer.

4.4.3 Fluorimetry

Hori *et al* have developed a new fluorimetric analysis for citric acid [57].

4.5 Chromatographic Methods of Analysis

4.5.1 Thin Layer Chromatography

A report describes the quantitative separation of citric acid from trichloroacetic acid on impregnated calcium sulfate plates using ethyl acetate as eluent and 1% ethanolic alkaline bromphenol blue solution as the detection reagent [58].

Citric acid can also be determined chromatographically using silica gel plates with several solvent systems. It has relatively low Rf values compared with the other dicarboxylic acids [59]. Some TLC systems for citric acid are presented in Table 16.

4.5.2 Gas Chromatography

Rosenqvist *et al* has reported the gas chromatographic analysis of citric acid cycle and related compounds from *Escherichia coli* as their trimethylsilyl derivatives [60].

Table 16

Thin Layer Chromatographic Methods for the Determination of Citric acid

Layer	Mobile phase	Developing Distance	Rf × 100 Values	Note
Silica gel G	96% Ethanol/ water/25% NH_4OH (100/12/16)	10 cm	5	Citric acid as the ammonium salt. Unsaturated atmosphere.
Silica gel G	Benzene/ methanol/acetic acid (45/8/4)	10 cm	2	-
Silica gel G	Benzene/ dioxane/acetic acid (90/25/4)	10 cm	2	-
Silica gel G/ Kieselguhr G (1/1)	Benzene/ ethanol/NH_4OH (10/20/5)	6 cm	4	-
Silica gel G/ Kieselguhr G (1/1)	Butyl acetate/ methanol/ NH_4OH (15/20/5)	6 cm	12	-
Silica gel G/ Kieselguhr G (1/1)	Butyl acetate/ acetic acid/water (30/20/10)	6 cm	49	-

4.5.3 High Performance Liquid Chromatography

A food analysis handbook cited that citric acid can be determined by HPLC using reverse phase ion exchange and on proprietary columns for organic acids [56]. There is also an issue [61] that describes a liquid chromatographic assay for citric acid in over the counter carbamide peroxide products that uses indirect UV detection.

The AOAC Official Method describes the determination of quinic, malic, and citric acids in cranberry juice cocktail and apple juice by two reverse phase liquid chromatography columns in tandem [55]. This procedure uses a system equipped with a Model U6K injector, Model 450 variable wavelength detector operating at 214 nm and 0.1 AUFS (Waters Associate, Inc.), and a computing integrator (Hewlett-Packard Integrator 3390 [replaced by No. 3396B], or equivalent). The separation of the acids is performed using Supelcosil LC-18, or equivalent {5 μm particle size, 25 cm × 4.6 mm} in tandem with and followed by a Waters Radial-Pak C18 cartridge {5 μm particle size, 10 cm long} used with a Radial Compression Module. A phosphate buffer of 0.2M KH_2PO_4 (adjusted to pH 2.4) is used as the mobile phase, and is eluted at 0.8 mL/min.

An assay for citric acid in an irrigation solution composed of citric acid, magnesium oxide, and sodium carbonate is mentioned in USP 24 [46]. The system uses a 0.2% v/v solution of sulfuric acid as the mobile phase, which is maintained at 40°C throughout the analysis. The liquid chromatograph is equipped with a refractive index detector and a 7.8 mm × 30 cm column that contains packing L17 (strong cation-exchange resin consisting of sulfonated cross-linked styrene-divinylbenzene copolymer in the hydrogen form, 7 - 11 μm in diameter). The column temperature is maintained at 40°C, and the flow rate is about 0.6 mL per minute.

Restek Corporation has developed a HPLC system that separates various carboxylic acids [62]. The system consists of a Ultra Aqueous C18 column having dimensions of 150 × 4.6 mm, and a 210 nm UV detector. A mobile phase mixture of 50 mM potassium phosphate adjusted to pH 2.5 and acetonitrile (99:1) at a flow rate of 1.5 mL/min. and a temperature of 25°C was used. The method was used for the separation of malonic acid, lactic acid, acetic acid, citric acid, succinic acid, and fumaric acid.

Other methods for the HPLC analysis of citric acid, as provided by various column manufacturers [63,64], are presented in Table 17.

Table 17

High Performance Liquid Chromatographic Methods for the Separation of Citric acid and Other Substances

Column	Flow rate (Temp.)	Detection	Mobile phase	Analytes / Matrices	Ref.
Water Spherisorb® C6, 5μm, 250 × 4.6mm	0.8 mL/min.	-	0.1 M KH_2PO_4, pH 2.0 w/H_3PO_4	Tartaric acid, malic acid, lactic acid, citric acid, succinic acid	63
Water Spherisorb® C8, 5μm, 250 × 4.6mm	0.8 mL/min.	-	0.2 M Phosphoric acid	Tartaric acid, lactic acid, malic acid, formic acid, acetic acid, citric acid, succinic acid, fumaric acid	63
Hamilton PRP-X300, 7μ, 150 × 4.1mm	2 mL/min.	UV 210 nm	0.001 N Sulfuric acid	Tartaric acid, malic acid, citric acid, lactic acid, acetic acid, succinic acid	63
Altech Organic Acid IOA-1000, 300 × 7.8mm	0.4 mL/min. (42°C)	RI (16×)	0.009 N Sulfuric acid	*cis*-Aconitic acid, oxaloacetic acid, citric acid, α-ketoglutaric acid, pyruvic acid, malic acid, lactic acid, succinic acid, fumaric acid in Krebs TCA Cycle	63

Table 17 (continued)

High Performance Liquid Chromatographic Methods for the Separation of Citric acid and Other Substances

Column	Flow rate (Temp.)	Detection	Mobile phase	Analytes / Matrices	Ref.
Altech Organic Acid IOA-1000, 300 × 7.8mm	0.6 mL/min. (30°C)	RI (16×)	0.01 N Sulfuric acid	Sucrose, glucose, citric acid, fructose, tartaric acid, malic acid, glycerol, acetic acid, lactic acid, methanol, ethanol	63
Altech Organic Acid OA-1000, 300 × 6.5mm	0.8 mL/min. (65°C)	UV 210 nm	0.01 N Sulfuric acid	Citric acid, tartaric acid, malic acid, succinic acid, lactic acid, acetic acid, propionic acid	63
Polypore H, 220 × 4.6mm	0.15 mL/min. (25°C)	UV 210 nm	0.01 N Sulfuric acid	Oxalic acid, citric acid, α-ketoglutaric acid, malic acid, succinic acid, formic acid, acetic acid, fumaric acid, butyric acid	63

Table 17 (continued)

High Performance Liquid Chromatographic Methods for the Separation of Citric acid and Other Substances

Column	Flow rate (Temp.)	Detection	Mobile phase	Analytes / Matrices	Ref.
Jordi Sulfonated DVB, 500Å, 500 × 10mm	1.5 mL/min.	RI @ 4× & UV 210 nm	0.01 M Phosphoric acid, pH 3/w NaOH	Oxalic acid, tartaric acid, citric acid, malic acid, formic acid, lactic acid, succinic acid, acetic acid, fumaric acid	63
Nucleosil® C18 AB, 5μm, 250 × 4.6mm	0.8 mL/min.	UV 220 nm	Water + 0.2% H_3PO_4	Tartaric acid, malonic acid, maleic acid, citric acid, succinic acid, acrylic acid	63
Hypersil® 100 C18, 5μm, 100 × 4.6mm	0.5 mL/min.	UV 210 nm	Methanol/ 0.2 M phosphoric acid (10/90)	Glutamic acid, glycolic acid, malic acid, citric acid, fumaric acid	63
alphaBond™ C18, 10μm, 300 × 3.9mm	0.7 mL/min. (25°C)	UV 215 nm	0.05 M potassium phosphate, pH to 2.5 with H_3PO_4	Oxalic acid, tartaric acid, malic acid, ascorbic acid, citric acid, succinic acid	63
Alltima™ C18-LL, 5μm, 150 × 4.6mm	0.7 mL/min. (30°C)	UV 220 nm	pH 2.5, 0.05 M Monobasic potassium phosphate	Tartaric acid, lactic acid, citric acid	63

Table 17 (continued)

High Performance Liquid Chromatographic Methods for the Separation of Citric acid and Other Substances

Column	Flow rate (Temp.)	Detection	Mobile phase	Analytes / Matrices	Ref.
Benson SS-100 H + Carbohydrate Column, 300 × 7.8mm	0.4 mL/min. (60°C)	RI	2.5 mM H_2SO_4	Citric acid, tartaric acid, glucose, malic acid, fructose, succinic acid, lactic acid, acetic acid, 2,3-butanediol, ethanol	63
ProntoSIL C18 AQ, 3μ, 250 × 3.1 mm	0.7 mL/min. (30°C)	UV 205 nm	0.05 M H_3PO_4	Glutamic acid, oxalic acid, tartaric acid, malic acid, ascorbic acid, acetic acid, maleic acid, citric acid, fumaric acid	63
Alltima™ 3μm C18 Rocket, 33 × 7 mm	2.3 mL/min.	UV 215 nm	0.05 M KH_2PO_4, pH 2.5	Oxalic acid, tartaric acid, malic acid, ascorbic acid, citric acid, succinic acid	63
Shodex SUGAR C-811 × 2	0.9 mL/min. (60°C)	Shodex RI & VD 430 nm	3 mM $HClO_4$	Citric acid, malic acid, glucose, fructose in orange juice	64

Table 17 (continued)

High Performance Liquid Chromatographic Methods for the Separation of Citric acid and Other Substances

Column	Flow rate (Temp.)	Detection	Mobile phase	Analytes / Matrices	Ref.
Shodex SUGAR SH 1011	1.0 mL/min. (15°C using Shodex Oven AO-30C)	Shodex RI	3 mM $HClO_4$	Sucrose, citric acid, glucose, fructose, lactic acid, succinic acid, formic acid, acetic acid	64
Shodex SUGAR SH 1011	1.0 mL/min. (15°C using Shodex Oven AO-30C)	Shodex RI	3 mM $HClO_4$	Stachyose, raffinose, sucrose, citric acid, glucose, fructose, succinic acid in soft drink	64
Shodex RSpak KC-G + KC-811 × 2	1.0 mL/min. (50°C)	Shodex OA	3 mM $HClO_4$	Citric acid, pyruvic acid, malic acid, succinic acid, lactic acid, formic acid, acetic acid, pyroglutamic acid	64
Shodex RSpak KC-811 × 4	1.0 mL/min. (40°C)	Shodex VD 430 nm	1 mM $HClO_4$	Oxalic acid, α-ketoglutaric acid, pyruvic acid, 2-ketoglutaric acid, citric acid, tartaric acid, malonic acid, etc.	64

Table 17 (continued)

High Performance Liquid Chromatographic Methods for the Separation of Citric acid and Other Substances

Column	Flow rate (Temp.)	Detection	Mobile phase	Analytes / Matrices	Ref.
Shodex RSpak KC-LG + KC-811 × 2	1.0 mL/min. (40°C)	Shodex VD 430 nm	3 mM $HClO_4$	α-Ketoglutaric acid, 2-ketoglutaric acid, citric acid, tartaric acid, gluconic acid, etc.	64
Shodex RSpak KC-811 × 2	1.0 mL/min. (50°C)	Shodex VD 430 nm	6 mM $HClO_4$	Citric acid, tartaric acid, malic acid, succinic acid, glycolic acid, lactic acid, etc.	64
Shodex RSpak KC-811	1.0 mL/min. (60°C)	-	0.1% H_3PO_4	Oxalic acid, citric acid, tartaric acid, malonic acid, succinic acid, etc.	64
Shodex RSpak KC-LG + DE-613 × 2 + KC-811 × 3	1.0 mL/min. (50°C)	Shodex VD 430 nm	2 mM $HClO_4$	Phosphoric acid, citric acid, malic acid.	64
Shodex RSpak KC-LG + KC-811 × 2	1.0 mL/min. (63°C)	Shodex VD 430 nm	4.8 mM $HClO_4$	Citric acid, pyruvic acid, gluconic acid, malic acid, in beer.	64

Table 17 (continued)

High Performance Liquid Chromatographic Methods for the Separation of Citric acid and Other Substances

Column	Flow rate (Temp.)	Detection	Mobile phase	Analytes / Matrices	Ref.
Shodex RSpak KC-LG + KC-811 × 2	1.0 mL/min. (40°C)	Shodex VD 430 nm	1 mM $HClO_4$	Anions (Cl^-), oxalic acid, maleic acid, citric acid, malic acid, vitamin C, succinic acid, formic acid, acetic acid.	64
Shodex RSpak KC-LG + KC-811 × 2	1.0 mL/min. (63°C)	Shodex VD 430 nm	4.8 mM $HClO_4$	Phosphoric acid, citric acid, pyruvic acid, malic acid, succinic acid, etc. in sake (Japanese rice wine).	64
Shodex RSpak KC-LG + KC-811 × 2	1.0 mL/min. (45°C)	Shodex VD 430 nm	3 mM $HClO_4$	Citric acid, tartaric acid, malic acid, succinic acid, lactic acid, etc. in white wine.	64
Shodex RSpak KC-811 × 2	1.0 mL/min. (50°C)	Shodex RI	50 mM $HClO_4$	Anions (Cl^-), oxalic acid, citric acid, tartaric acid, phophoric acid, etc.	64

Table 17 (continued)

High Performance Liquid Chromatographic Methods for the Separation of Citric acid and Other Substances

Column	Flow rate (Temp.)	Detection	Mobile phase	Analytes / Matrices	Ref.
Shodex RSpak KC-811 × 2	1.0 mL/min. (50°C)	Shodex UV 210 nm	1 mM $HClO_4$	Citric acid, malic acid, vitamin C, succinic acid	64
Shodex IC I-524A	1.2 mL/min. (40°C)	Shodex CD	1.5 mM Phthalic acid + 1.38 mM tris(hydroxymethyl)aminomethane + 300 mM boric acid (pH 4.0)	Acetic acid, $H_2PO_4^-$, succinic acid, pyroglutamic acid, lactic acid, pyruvic acid, Cl^-, malic acid, NO_3^-, oxalic acid, citric acid, SO_4^{2-}, tartaric acid in beer	64
Shodex RSpak KC-811	1 mL/min. (60°C)	Shodex CD & UV 210 nm	3 mM $HClO_4$	Citric acid, malonic acid, succinic acid, lactic acid, formic acid, acetic acid	64

4.5.4 Ion Chromatography

Ion chromatographic methods [63] have been developed to separate citric acid and other substances present in various complex mixtures using specific columns and conductivity detection. The details of these methods are collected in Table 18.

4.5.5 Capillary Electrophoresis

Application of the capillary electrophoresis (CE) technique permitted a tremendous flexibility in separation selectivity and rapid micro-volume analysis for the separation of small organic molecules in complex matrices such as fermentation broths or botanical extracts. Citric acid is included in a group of organic acids that can be appropriately analyzed using the CE method [65].

More recently, a method of capillary electrophoresis has been developed, optimized, and validated for measuring of citric, isocitric, malic, and tartaric acids as authenticity markers in orange juices. The method is run without any sample treatment, other than dilution and filtration. The method conditions were an eluent of phosphate buffer (200 mM, pH 7.50), an applied potential of – 14 kV, and a neutral capillary length of 57 cm. Detection was by direct UV absorbance at 200 nm [66].

4.5.6 Column Chromatography

An experimental organic chemistry handbook reveals a simple column chromatographic procedure that can be used to separate a mixture of citric acid and glucose by base anion resin [67]. The system uses strong base anion exchange resin for packing, and this is packed in a column measuring 75 × 1.5 cm glass tube which is constricted at one end. Sodium carbonate solution (1 N)is used to wash the column, and while citric acid is retained by the column, glucose elutes first. The column is washed with distilled water until all of the glucose is eluted out. Citric acid can be recovered from the column by eluting with a 1 N solution of ammonium carbonate.

Table 18

Ion Chromatographic Methods for the Separation of Citric acid and Other Substances

Column	Flow rate (Temp.)	Detection	Mobile phase	Analytes / Matrices
Wescan Anion Exclusion, 100 × 7.5 mm	0.8 mL/min.	Conductivity	1 mM Sulfuric acid	Citric acid, gluconic acid, succinic acid, formic acid, acetic acid, adipic acid
Jordi Organic Acid Column, 250 × 10 mm	1.0 mL/min. (35°C)	Conductivity	0.5 mM Benzoic acid in 10% acetonitrile	Oxalic acid, citric acid, formic acid, succinic acid, acetic acid, propionic acid, butyric acid
Jordi Organic Acid Column, 250 × 10 mm	1.0 mL/min. (35°C)	Conductivity	3 mM Sulfuric acid	Ascorbic acid, citric acid in grape fruit juice
Jordi Organic Acid Column, 250 × 10 mm	1.0 mL/min. (35°C)	Conductivity	0.5 mM Benzoic acid in 10% acetonitrile	Citric acid, ascorbic acid, acetic acid in multivitamin
Jordi Organic Acid Column, 250 × 10 mm	1.0 mL/min.	Conductivity	3 mM Sulfuric acid	Oxalic acid, tartaric acid, citric acid, maleic acid

4.6 Gravimetric Method of Analysis

This method is also called as the pentabromoacetone method, since the citric acid is oxidized by permanganate to acetonedicarboxylic acid, which then reacts with bromine to form pentabromoacetone. The collected pentabromoacetone precipitate is dried overnight in a desiccator and weighed [55].

It has been observed that the pentabromoacetone precipitate can also be measured colorimetrically [6].

4.7 Determination in Body Fluids and Tissues

It was recommended that the determination of citric acid in blood and other biological materials could be readily accomplished by its conversion to pentabromoacetone, which is then measured either gravimetrically or colorimetrically [6].

Oefner *et al* reported the application of a HPLC method for the analysis of citric acid, inositol, and fructose in boar and bull seminal plasma [68].

A simultaneous determination of oxalate, glycolate, citrate, and sulfate from dried urine filter paper spots in a pediatric population was issued by Blau *et al* [69].

5. Stability

5.1 Solid-State Stability

Citric acid is both an α- and β-hydroxy acid, and exhibits the characteristic reactions of each. When heated to 175°C, it is partially converted to aconitic acid by elimination of water, and to acetonedicarboxylic acid by the loss of carbon dioxide and water. At temperature above 175°C, citric acid yields an oily distillate, which crystallizes as itaconic acid. Further heating yields a non-crystallizable oil which is citraconic anhydride. The hydrogenation of citric acid yields tricarballylic acid.

Digestion of citric acid with fuming sulfuric acid, or oxidation with potassium permanganate, solution yields acetonedicarboxylic acid, a reaction which is characteristic of α-hydroxy acids. When this is heated more strongly with concentrated sulfuric acid, it decomposes to acetone and two molecules of carbon dioxide. Above 35°C, oxidation with potassium permanganate produces oxalic acid. Citric acid decomposes to form oxalic acid and acetic acid when fused with potassium hydroxide or when oxidized with nitric acid [3,6,29].

5.2 Solution-Phase Stability

In aqueous solution, citric acid can be mildly corrosive toward carbon steels, and should therefore be used with an appropriate inhibitor. It is not corrosive to the stainless steels that are most often employed as the material of construction for processes involving citric acid.

At proper pH in aqueous media, the hydroxyl and carboxylic acid groups of citric acid act as multidentate ligands, forming complexes or chelates with metal ions. These chelating reactions are the basis for many industrial processes, including elimination or control of metal-ion catalysis, lowering of metal oxidation potentials, removal of corrosion products (i.e., Fe^{3+}), regeneration of ion-exchange resins, recovery of valuable metals by precipitation of insoluble chelates, decontamination of radio-active materials, quenching reactions, and driving reactions to completion [6].

Dilute aqueous solutions of citric acid may ferment on standing [2,4].

5.3 Stability in Biological Fluids

Citric acid is found naturally in the body, mainly in the bones, and in small amounts in blood and in urine [4,29]. The citrate ion occurs in all human tissues and fluids [6], such as in human whole blood (15 ppm), blood plasma (25 ppm), red blood cells (10 ppm), milk (500-1250 ppm), urine (100-750 ppm), semen (2000-4000 ppm), cerebrospinal fluid (25-50 ppm), mammary gland (3000 ppm), thyroid gland (750-900 ppm), kidney (20 ppm), bone (7500 ppm), amniotic fluid (17-100 ppm), saliva (4-24 ppm), sweat (1-2 ppm), tears (5-7 ppm), and skeletal muscle, liver and brain (2-100 ppm).

5.4 Incompatibilities with Functional Groups

Citric acid yields a greenish color with ferric chloride; effervesces with alkali and alkaline earth carbonates and bicarbonates, and releases hydrogen sulfide from soluble sulfides. The substance is incompatible with potassium tartrate and acetates. Other incompatibilities also include oxidizing agents, bases, reducing agents, and nitrates. Citric acid may be potentially explosive in combination with metal nitrates. On storage, sucrose may crystallize from syrups in the presence of citric acid [2,4,41].

6. Drug Metabolism and Pharmacokinetics

Citric acid is a normal metabolite and an intermediate in cellular oxidative metabolism. It is formed in the mitochondrion after condensation of acetate with oxaloacetate. The six-carbon acid is then successively degraded to a series of four-carbon acids, effectively accomplishing the oxidation of acetate in the cell [5]. Thus, citric acid is metabolized to carbon dioxide and water, and has only a transient effect on systemic acid-base status; it works as a temporary buffer component. Oxidation is virtually complete, and less than 5% of citrates are excreted in urine unchanged [79].

Citric acid is commonly consumed as part of a normal diet. Man's total daily consumption of citric acid from natural sources and from food additive sources may exceed 500 mg/kg of body weight [4,5]. Citric acid is well absorbed from the gastrointestinal tract, and reacts with the enzyme citratase to yield oxaloacetic acid and acetic acid [27,70]. The total circulating citric acid in the serum of a man is approximately 1 mg/kg of body weight. Normal daily excretion in the urine of humans is 0.2-1.0 g [6].

Orally ingested citric acid is absorbed, and is generally regarded as a nontoxic material when used as an excipient. However, excessive or frequent consumption of citric acid has been associated with erosion of the teeth [1,4,5,71,72].

Citric acid (5 percent in the diet) did not depress food intake, but caused a loss in body weight gain and survival time in mice, with a slightly greater influence on mature animals. The effects on body weight gain and survival time may have resulted from the chelating ability of citric acid, which could impair absorption of calcium and iron. The use of large volumes of blood anti-coagulated with citrates has been shown to decrease plasma levels of calcium and magnesium [5,20].

Citric acid and citrates also enhance intestinal aluminum absorption in renal patients that may lead to increased and harmful serum aluminum levels. Although aluminum is an abundant terrestrial element, it is toxic to tissues, including the brain. It has therefore been suggested that patients with renal failure taking aluminum compounds to control phosphate absorption should not be prescribed citric acid and or citrate-containing products [1,4,73].

A study by Lacour *et al* indicates that citric acid supplementation together with a Ca-rich diet allows an increased retention of calcium and phosphorous in bone [74]. Addition of citric acid supplements to a maize-soya diet enhanced the dialysability of calcium, magnesium, manganese and zinc. A a dose-dependent increase of lead and cadmium dialysability due to citric acid supplementation was evident [75].

The result of a kinetic assessment of salivary secretory response to citric acid is citric acid provoked a rapid and short-lived salivary response that differs markedly from the one produced by other secretagogues, such as pilocarpine [76].

Citric acid is moderately strong acid with some irritant and allergenic properties. A splash of a saturated citric acid solution in human eyes caused severe conjunctival reaction and corneal ulceration. Application of 500 mg citric acid to rabbit skin produced a moderate irritation in 24 hours, whereas 750 μg causes severe effects in rabbit eye [4,5]. The acute toxicity of citric acid is summarized below:

Species	Route and LD_{50} (mg/kg)			
	Iv	Oral	Ip	Sc
Mouse	42	5040	961	2700
Rat	-	11700	883	5500
Rabbit	330	7000 (LD_{Lo})	-	-

In the gastrointestinal system, sodium citrate and citric acid solutions have a direct irritating effect on oral mucous membranes, and may cause necrotic and ulcerative lesions [20].

The aerosol spray of citric acid stimulates the tracheal sensations produced by cigarette smoke and satisfies smokers' desire for cigarettes. Smokers rated the respiratory tract sensations produced by citric acid aerosol equal to or better than that of low tar and nicotine cigarettes in terms of liking, similarity to their brand, and reduction in cigarette craving [79].

Citric acid may be a mild to moderate irritant if inhaled in an aerosol. After the exposure has been terminated, patients should be evaluated for respiratory distress [27,77]. A study of citric acid inhalation in guinea pigs concludes that citric acid-induced bronchoconstriction is caused by tachykinin release from sensory nerves, which in part, is mediated by endogenously released bradykinin. Simultaneous release of nitric oxide by citric acid inhalation counteracts tachykinin-mediated bronchoconstriction [78].

7. References

1. ***Martindale: The complete drug reference***, 32nd edn., K. Parfitt, ed., Pharmaceutical Press, Taunton, Massachusetts, 1999, pp. 1564-1565.

2. ***The Merck Index***, 12th edn., Merck Research Laboratories, Division of Merck & Co., Inc., Whitehouse Station, N.J., 1996, pp. 2384-2385.

3. I.L. Finar, ***Organic Chemistry Volume 1: The Fundamental Principles***, 6th edn, Longman Singapore Publishers (Pte) Ltd., Singapore, 1986, pp. 492-493.

4. ***Handbook of Pharmaceutical Excipients***, 3rd edn., A.H. Kibbe, ed., American Pharmaceutical Association and Pharmaceutical Press, Washington-London, 2000, pp. 140-142.

5. ***Patty's Industrial Hygiene and Toxicology***, 3rd revised edn., G.D. Clayton and F.E. Clayton, eds., Volume 2C: Toxicology, John Wiley & Sons, Inc., New York, 1982, pp. 4936-4937, 4946-4947.

6. R.E. Kirk, D.F. Othmer, M. Grayson, and D. Eckroth, ***Kirk-Othmer Encyclopedia of Chemical Technology***, 3rd edn., Volume 6, John Wiley & Sons, Inc., New York, 1979, pp. 150-179.

7. ***British Pharmacopoeia 2000***, Volume 1, The Stationery Office, Norwich, U.K., 2000, pp. 405-407.

8. ***Pharmacopoeia of the Peoples' Republic of China (English Edition 1997)***, Volume II, Chemical Industry Press, Beijing, 1997, p. 136.

9. ***Indian Pharmacopoeia 1996***, Volume I (A-O), Published by the Controller of Publications, New Delhi, 1996, pp. 192-193.

10. ***Farmakope Indonesia*** (Indonesian Pharmacopoeia), Edisi IV, Departemen Kesehatan Republic Indonesia, Jakarta, 1995, p. 48.

11. Anhydrous citric acid and citric acid monohydrate (European Pharmacopoeia), http://www.the stationary_office.co.uk/ bp2k/bpfrls.htm (Dec. 6, 2000).

12. ***Grant & Hackh's Chemical Dictionary***, 5th edn., R. Grant and C. Grant, eds., McGraw-Hill Book Company, New York, pp. 138-139.

13. ***Natural Products of Woody Plants I***, J.W. Rowe, ed., Springer-Verlag, Berlin Heidelberg, 1989, pp. 260-261.

14. I.L. Finar, ***Organic Chemistry Volume 2: Stereochemistry and the Chemistry of Natural Products***, 5th edn, Longman Singapore Publishers (Pte) Ltd., Singapore, 1988, pp. 689-690.

15. ***Patty's Industrial Hygiene and Toxicology***, 3rd edn., Volume III, Part B: Theory and Rationale of Industrial Hygiene Practice: Biological Responses, L.J. Cralley, L.V. Cralley, and J.S. Bus, eds., John Wiley & Sons, Inc., New York, 1995, p. 602.

16. K.A. Connors, G.L. Amidon, and V.J. Stella, ***Chemical Stability of Pharmaceuticals***, 2nd edn., John Wiley & Sons, Inc., New York, 1986, p. 100.

17. U. Rosenberger, M. Shakibaei, C. Weise, P. Franke, and K. Buchner, *J. Cell. Biochem.*, **59**, 177 (1995).

18. T.A. Adesunloye and P.E. Stach, *Drug Dev. Ind. Pharm.*, **24**, 493 (1998).

19. B.Z. Jekone, *Acta Pharm. Hung.*, **68**, 133 (1998).

20. ***The Pharmacological Basis of Therapeutics***, 5th edn., L.S. Goodman and A. Gilman, eds., Macmillan Publishing Co., Inc., New York, 1975, p. 806.

21. ***Remington's Pharmaceutical Sciences***, 15th edn., A.Osol and J.E. Hoover, eds., Mack Publishing Company, Easton, PA, 1975.

22. N.W. Levin, S.L. Parnell, N.H. Prince, F. Gotch, H.D. Polaschegg, R. Levin, A. Alto, and A.M. Kaufman, *J. Am. Soc. Nephrol.*, **6**, 1578 (1995).

23. B.S. Nagoba, S.R. Desmukh, B.J. Wadher, L. Mahabaleshwar, R.C. Gandhi, P.B. Kulkarni, V.A. Mane, and J.S. Desmukh, *J. Hosp. Infect.*, **40**, 155 (1998).

24. Abstract of CEH report: Citric acid, http://ceh.sric.sri.com/Public/Reports/636.5000/Abstract.html (Dec.11, 2000).

25. Purchasing, http://www.findarticles.com/cl_1/m3148/9_128/62601779/print.jhtml (Dec. 12, 2000).

26. Organic acids/Market share, http://www.findarticles.com/cl_1/mOFVP/22_257/62743149/p2/article.jhtml (Dec. 12, 2000).

27. MICROMEDEX(R)Healthcare Series: Citric acid, file://A:\MICROMEDEX(R)Healtcare SeriesCitric Acid.htm (March 22, 2001).

28. B. Hoppe, B. Roth, C. Bauerfeld, and C.B. Langman, *J. Pediatr. Gastroenterol. Nutr.*, **27**, 383 (1998).

29. W. Walter, ***Beyer-Walter Handbook of Organic Chemistry***, Translated by D. Lloyd, Prentice Hall Europe, London, 1996, pp. 349-350.

30. Citric acid: Producer and Capacity, http://www.findarticles.com/cl_1/mOFVP/n6_V254/21007493/p1/article.jhtml (Dec. 18, 2000).

31. Citric Acid Production from Concentrated Milk-wastewater, http://chem.pknu.ac.kr/~biolab/paper/79.htm (Dec. 6, 2000).

32. S. Mourya and K.S. Jauhri, *Microbiol. Res.*, **155**, 37 (2000).

33. G.M. Gadd, *Adv. Microb. Physiol.*, **41**, 47 (1999).

34. K. Kirimura, T. Watanabe, T. Sunagawa, and S. Usami, *Biosci. Biotechnol. Biochem.*, **63**, 226 (1999).

35. F.H. Verhoff and J.E. Spradlin, *Biotechnol. Bioeng.*, **18**, 452 (1976).

36. Fuyang Pharmacy, http://www.hunanjiali.com/html/factory.htm (Dec. 8, 2000).

37. BBCA Biochemical Group Corp., http://www.chinadaily.com.cn/sup/domes/d092832.htm (Dec. 11, 2000).

38. Nantong Huaze Chemical Co., Ltd., http://www.chinacitricacid.com/products2.htm (Dec. 6, 2000).

39. Citric acid/Supply and demand, http://www.findarticles.com/cl_1/m3148/9_125/53604166/p1/article.jhtml (Dec. 18, 2000).

40. R.J. Tiemko and N.G. Lordi, *J. Pharm. Sci.*, **68**, 601 (1979).

41. ***Dispensing of Medication***, 8th edn., J.E. Hoover, ed., Mack Publishing Company, Easton, PA, 1976, p. 504.

42. ***CRC Handbook of Chemistry and Physics***, 70th edn., R.C. Weast, ed.-in-chief, CRC Press, Inc., Boca Raton, Florida, 1990, pp. D-165, D-228.

43. ***Lange's Handbook of Chemistry***, 11th edn., J.A. Dean, ed., McGraw-Hill Book Company, New York, 1973, pp. 5.21, 5.33, 10.91.

44. ***Perry's Chemical Engineers' Handbook***, 7th edn., D.W. Green and J.O. Maloney, eds., McGraw-Hill Co., Inc.,New York, 1997, p. 2.33.

45. R.L. Shriner, R.C. Fuson, D.Y. Curtin, and T.C. Morrill, ***The Systematic Identification of Organic Compounds, a laboratory manual***, 6th edn., John Wiley & Sons, Inc., New York, 1980, p. 534.

46. ***United States Pharmacopoeia 24*** (Asian Edition), Unites States Pharmacopoeial Convention, Inc., Rockville, MD, 1993, p. 423.

47. D.C. Garratt, ***The Quantitative Analysis of Drugs***, 3rd edn., Chapman & Hall Ltd., Toppan Company Ltd., London-Tokyo, 1964, pp. 182-183.

48. K. Vytras, "Contemporary Trends in the Use of Ion-Selective Electrodes in the Analysis of Organic Substances", chapter 7 in ***Advanced Instrumental Methods of Chemical Analysis***, J. Churácek, P.J. Cox, M.R. Masson, eds., Ellis Horwood Ltd., New York, 1993, pp. 142-164.

49. S. Mlinko, *Mikrochim. Acta*, **1963**, 456 (1963).

50. M. Noshino and T. Yarita, *Bunseki Kagaku*, **24**, 390 (1975).

51. U. Fiedler, E. Hansen, and J. Ruzicka, *Anal. Chim. Acta*, **74**, 423 (1975).

52. L. Begheun, *Analyst*, **101**, 710 (1976).

53. P.W. Alexander, P.R. Haddad, and M. Trojanowicz, *Anal. Chim. Acta*, **171**, 151 (1985).

54. M.I. Prodromidis, S.M. Tzouwara-Karayanni, M.I. Karayannis, and P.M. Vadgama, *Analyst*, **122**, 1101 (1997).

55. ***Official Methods of Analysis of AOAC International***, 16th edn., Volume II: Food, Additives, Natural Contaminants, P. Cunniff, ed., AOAC International, Arlington, Virginia, 1995, pp. 28:10, 33:7-8, 65, 37:13-14.

56. R.S. Kirk and R. Sawyer, ***Pearson's Composition and Analysis of Foods***, 9th edn., Longman Singapore Publishers (Pte) Ltd., Singapore, 1991, p. 542.

57. M. Hori, T. Kometani, H. Ueno, and H. Morimoto, *Biochem. Med.*, **11**, 49 (1974).

58. H. Rathore, K. Kumari, and M. Agrawal, *J. Liquid Chromatogr.*, **8**, 1299 (1985).

59. J. G. Kirchner, ***Thin layer Chromatography***, 2nd edn., Sigma Chemical Company, 1990, p. 360.

60. H. Rosenqvist, H. Kallio, and V. Nurmikko, *Anal. Biochem.*, **46**, 224 (1972).

61. T.A. Walker, *J. Pharm. Biomed. Anal.*, **13**, 171 (1995).

62. Restek Advantage Chromatography Newsletter, http://www.restekcorp.com/advantage/figd00-/.htm (Dec.8, 2000).

63. Altech's ChromAccess Online, http://www.altechweb.com/chromaccess/hplcchromaccess (Dec. 8, 2000).

64. High-selective Analysis of Organic Acids, http://www.hplc1.com/shodex/english/ic41.htm (Dec. 7, 2000).

65. Capillary Electrophoresis @ Alpha Labs, http://www.alphalabs.com/ce.html (Dec. 8, 2000).

66. L. Saavedra, A. Garcia, and C. Barbas, *J. Chromatogr. A*, **881**, 395 (2000).

67. P.R. Singh, D.S. Gupta, and K.S. Bajpal, ***Experimental Organic Chemistry***, Volume 1, Tata McGraw-Hill Publishing Company, Ltd., New Delhi, 1980, pp. 70-71.

68. P. Oefner, G. Bonn, and G. Bartsch, *Andrologia*, **17**, 250, (1985).

69. N. Blau, A. Matasovic, A. Lukasiewicz-Wedlechowicz, C.W. Heizmann, and E. Leumann, *Clin. Chem.*, **44**, 1554 (1998).

70. ***Fenaroli's Handbook of Flavor Ingredients***, Volume 2, 2nd edn., T.E. Furia and N. Bellanca, eds., The Chemical Rubber Company, Cleveland, OH, 1975.

71. J.A. Hughes, N.X. West, D.M. Parker, M.H. van den Braak, and M. Addy, *J. Dent.*, **28**, 147 (2000).

72. M.W. Dodds, P.P. Gragg, and D. Rodriquez, *Pediatr. Dent.*, **19**, 339 (1997).

73. G.A. Taylor, P.B. Moore, I.N. Ferrier, S.P. Tyrer, and J.A. Edwardson, *J. Inorg. Biochem.*, **69**, 165 (1998).

74. B. Lacour, S. Tardivel, and T. Drueke, *Miner. Electrolyte Metab.*, **23**, 79 (1997).

75. A. Walter, G. Rimbach, E. Most, and J. Pallauf, *Zentralbl. Veterinarmed. A*, **45**, 517 (1998).

76. V. Duran, P. Dominguez, I. Morales, and R.O. Lopez, *Rev. Med. Chil.*, **126**, 1330 (1998).

77. T. Kondo, I. Kobayashi, N. Hayama, and Y. Ohta, *Jpn. J. Physiol.*, **48**, 341 (1998).

78. F.L.M.Ricciardolo, V. Rado, L.M. Fabbri, P.J. Sterk, G.U. Di Maria, and P. Geppetti, *Am. J. Respir. Crit. Care Med.*, **159**, 557 (1999).

79. MICROMEDEX(R)Healthcare Series: Citrate Salts, file://A:\MICROMEDEX(R)Healtcare SeriesCitric2DrugDex.htm (March 22, 2001).

80. ***Critical Stability Constants***, volume 3, A.E. Martell and R.M. Smith, Plenum Press, New York, 1977, pp. 160-164.

FLAVOXATE HYDROCHLORIDE

Yuri Goldberg

Apotex Inc.
150 Signet Drive
Toronto, Ontario M9L 1T9
Canada

1075-6280/01 $35.00

Contents

1. Description

1.1 Nomenclature

1.1.1 Systematic Chemical Names

2-(1-Piperidino)ethyl 3-methyl-4-oxo-2-phenyl-4H-1-benzopyran-8-carboxylate hydrochloride

2-(1-Piperidino)ethyl 3-methylflavone-8-carboxylate hydrochloride

2-(1-Piperidino)ethyl ester of 3-methyl-4-oxo-2-phenyl-4H-1-benzopyran-8-carboxylic acid, hydrochloride

1-Piperidineethanol, 3-methyl-4-oxo-2-phenyl-4H-1-benzopyran-8-carboxylate, hydrochloride

1.1.2 Nonproprietary Names

Flavoxate Hydrochloride

1.1.3 Proprietary Names [1, 2]

Urispas, Bladderon, Genurin, Spasuret, Patricin

1.2 Formulae

1.2.1 Empirical Formula, Molecular Weight, CAS Number

1.2.1.1 Free base

$C_{24}H_{25}NO_4$ [MW = 391.47]

CAS number = 15301-69-6

1.2.1.2 Hydrochloride salt

$C_{24}H_{25}NO_4 \bullet HCl$ [MW = 427.96]

CAS number = 3717-88-2

1.2.2 Structural Formula

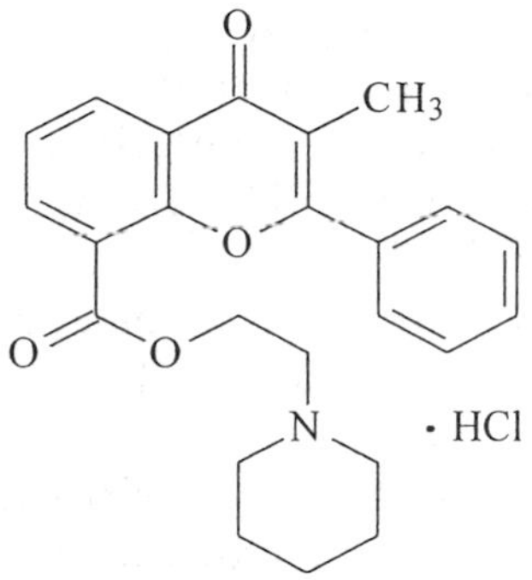

1.3 Elemental Analysis

The calculated elemental composition, and that obtained in this work, is as follows:

	Calculated for $C_{24}H_{25}NO_4 \bullet HCl$	Found
carbon:	67.36	67.34
hydrogen:	6.12	6.03
nitrogen:	3.27	3.16
chlorine:	8.28	8.06

1.4 Appearance

Flavoxate hydrochloride is a white or almost white crystalline powder [3].

1.5 Uses and Applications

Flavoxate hydrochloride is a synthetic urinary tract antispasmodic that exerts a direct spasmolytic effect on smooth muscle and provides therapeutic benefits in a variety of urological disorders. Flavoxate is applied in the symptomatic treatment of dysuria, urgency, nocturia, supraburic pain, frequency, and urge incontinence due to a variety of urological conditions [4-12].

2. Method(s) of Preparation [13]

The synthetic route for the preparation of Flavoxate HCl is shown in Scheme 1. The key intermediate, 3-methyl-2-phenylflavone-8-carboxylic acid (7), was obtained starting from 2-hydroxy-3-nitropropiophenone (1). Methylation of (1) with dimethyl sulfate, followed by reduction of intermediate (2) with iron in hydrochloric acid, furnished 3-amino-2-methoxypropiophenone (3). The latter was converted into the corresponding cyano derivative (4) by means of the Sandmeyer reaction. Cleavage of the methoxy group of (4) yielded the phenolic intermediate (5) suitable for the cyclization to the substituted flavone (6) by reaction with benzoyl chloride/sodium benzoate at elevated temperature (180-190°C). Acidic hydrolysis of (6) gave rise to the desirable derivative of flavone-8-carboxylic acid (7).

The latter method was employed to prepare a number of basic esters by reacting the corresponding acid chloride with a number of 2-(alkylamino)ethanols. When 2-(1-piperidino)ethanol was used, the reaction gave 2-(1-piperidino)ethyl 3-methyl-4-oxo-2-phenyl-4H-1-benzopyran-8-carboxylate hydrochloride (i.e., Flavoxate HCl). Alternatively, the intermediate cyano substututed flavone (4) can be obtained from 3-methyl-8-aminoflavone (readily obtainable from (1) [14]) by applying the Sandmeyer reaction [13].

3. Physical Properties

3.1 Ionization Constants

Flavoxate free base is characterized by a single ionization constant, for which the pKa was determined by potentiometric titration to be 7.3 (in water at 37°C) [15].

3.2 Solubility Characteristics

The solubilities of Flavoxate hydrochloride in different solvents are summarized in Table 1, and the pH dependence of the aqueous solubility is found in Table 2. Flavoxate free base is practically insoluble in water (0.001% w/v at 25°C), but soluble in common organic solvents (chloroform, methanol, ethanol, acetone and ethyl ether) [15, 16].

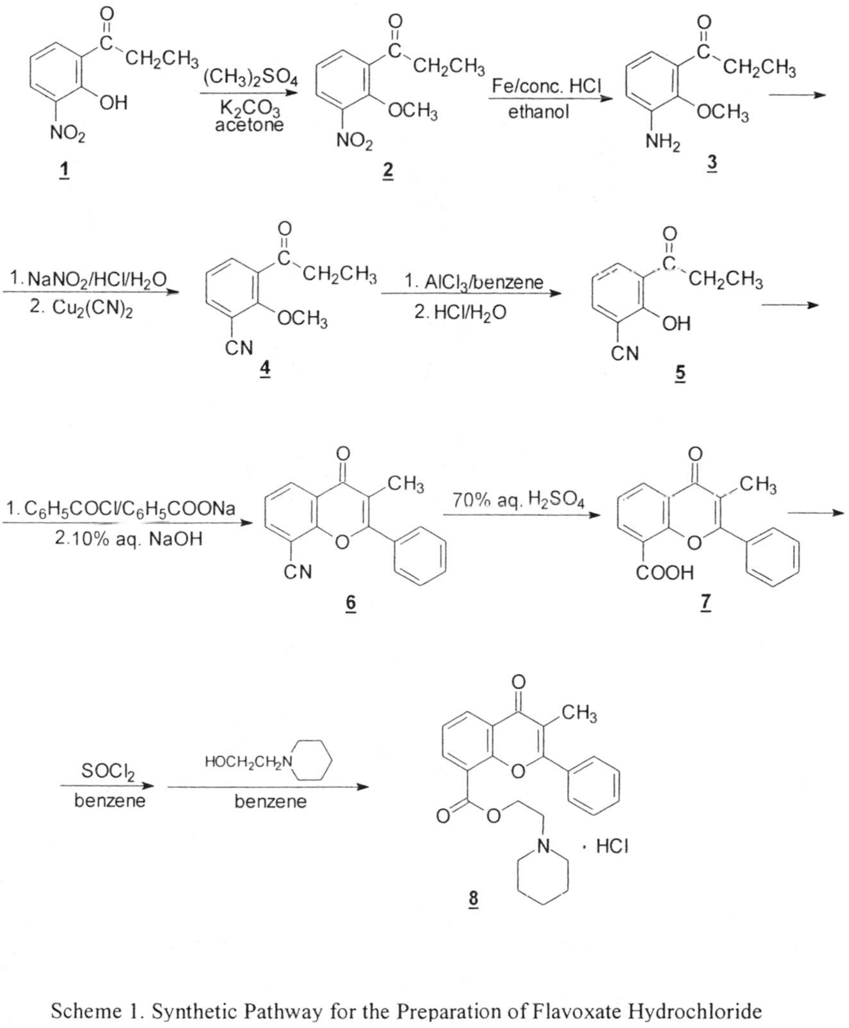

Scheme 1. Synthetic Pathway for the Preparation of Flavoxate Hydrochloride

Table 1

Solubility of Flavoxate Hydrochloride in Various Solvents at 20°C [15]

Solvent	Solubility (% w/v)
Water	0.75
Methanol	0.90
Ethanol	0.20
Chloroform	2.50
Acetone	Insoluble
Ethyl ether	Insoluble
n-Octanol	0.02

Table 2

pH Dependence of the Aqueous Solubility of Flavoxate Hydrochloride at Ambient Temperature

Medium	pH	Solubility (mg/mL)
Water	5.7	11.30
0.1N HCl	1.2	0.57
0.05N phosphate buffer	2.5	7.64
0.05N phosphate buffer	4.5	13.68
0.05N phosphate buffer	6.0	15.43
0.05N phosphate buffer	6.8	6.17
0.05N phosphate buffer	7.2	1.04
0.05N phosphate buffer	7.5	0.58

3.3 Partition Coefficients

The octanol / water partition coefficient of Flavoxate hydrochloride was found to be 2.01. The distribution coefficient between octanol and 0.1N HCl was found to be 4.51, and the distribution coefficient between octanol and pH 7.4 phosphate buffer was determined as 84.0 [15].

3.4 X-Ray Powder Diffraction Pattern

The x-ray powder diffraction pattern of Flavoxate hydrochloride was obtained on a Philips PW3710 diffractometer using Cu K_α radiation, and is shown in Figure 1. The sample exhibited a number of scattering peaks indicative of well-formed molecular planes in the crystals. A summary of the scattering angles, d-spacings, and relative intensities is provided in Table 3 (only peaks having relative intensities greater than or equal to 1% are listed).

3.5 Thermal Methods of analysis

3.5.1 Melting Behavior

In separate reports, Flavoxate hydrochloride has been found to melt with decomposition at 232-234°C [13], and at 230-233°C [15] (uncorrected). The melting point range obtained in the present work was 235.2-238.6°C (corrected).

3.5.2 Differential Scanning Calorimetry

The DSC thermogram for Flavoxate hydrochloride was obtained on a Perkin Elmer DSC 7 Series thermal analyzer, with the sample being heated at a rate of 10°C /min. The thermogram is shown in Figure 2, where a melting endotherm was observed with an onset of 243°C and maximum of 246°C. The thermogram shows that the melting is accompanied by decomposition.

3.5.3 Thermogravimetric Analysis

The thermal gravimetric analysis (TGA) was performed on a Perkin Elmer TGA 7 Series thermal analyzer using heating rate of 5°C /min. No weight loss was recorded between ambient temperature and 175°C, indicating that the sample contained neither moisture nor residual solvents.

Figure 1. X-ray powder diffraction pattern of Flavoxate hydrochloride.

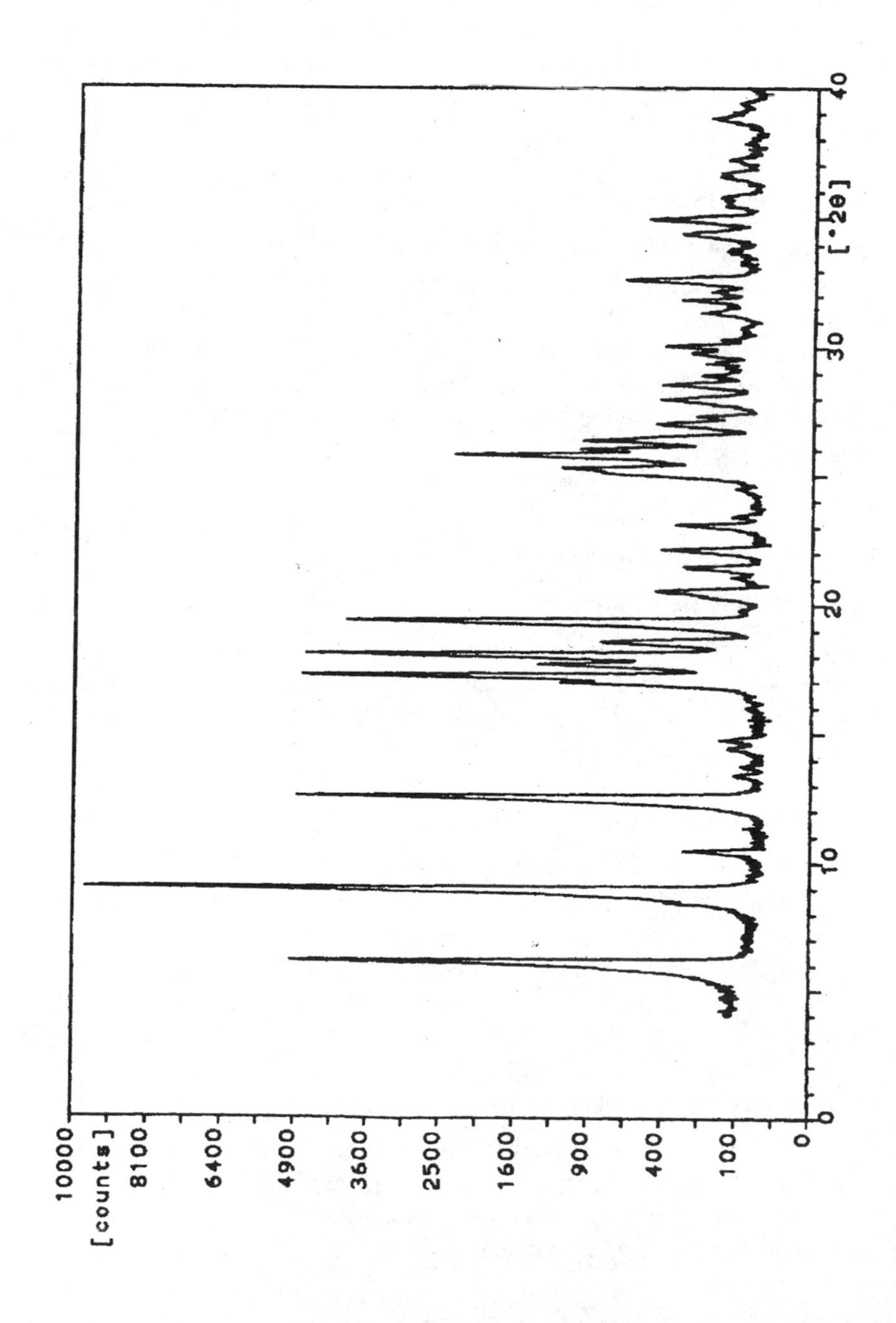

Table 3

Crystallographic Results from The X-Ray Powder Diffraction Pattern of Flavoxate Hydrochloride

Scattering Angle (degrees 2-)θ	d-Spacing (Å)	Relative Intensity (%)
6.170	14.3129	49.9
8.470	10.4307	3.2
9.005	9.8122	100.0
10.480	8.4342	2.5
12.570	7.0362	49.6
16.995	5.2128	11.3
17.270	5.1304	48.5
17.710	5.0040	13.8
18.095	4.8983	48.2
18.575	4.7728	7.5
19.385	4.5752	40.4
20.555	4.3173	4.1
21.485	4.1325	2.2
22.170	4.0063	3.8
23.090	3.8488	2.8
25.080	3.5477	7.1
25.280	3.5201	11.2
25.800	3.4503	23.8
26.035	3.4197	9.6
26.375	3.3764	9.2
26.995	3.3002	4.1
27.305	3.2635	2.0

Table 3 (continued)

Crystallographic Results from The X-Ray Powder Diffraction Pattern of Flavoxate Hydrochloride

27.970	3.1874	3.9
28.570	3.1218	3.8
28.935	3.0832	1.5
29.415	3.0340	1.3
29.800	2.9957	2.3
30.085	2.9979	3.7
31.380	2.8483	1.8
31.870	2.8056	2.7
32.205	2.7772	1.4
32.670	2.7387	6.2
34.400	2.6049	2.7
34.970	2.5637	4.6
35.795	2.5065	1.1
36.645	2.4503	1.1
38.820	2.3178	1.4

Figure 2. Differential scanning calorimetry thermogram of Flavoxate hydrochloride.

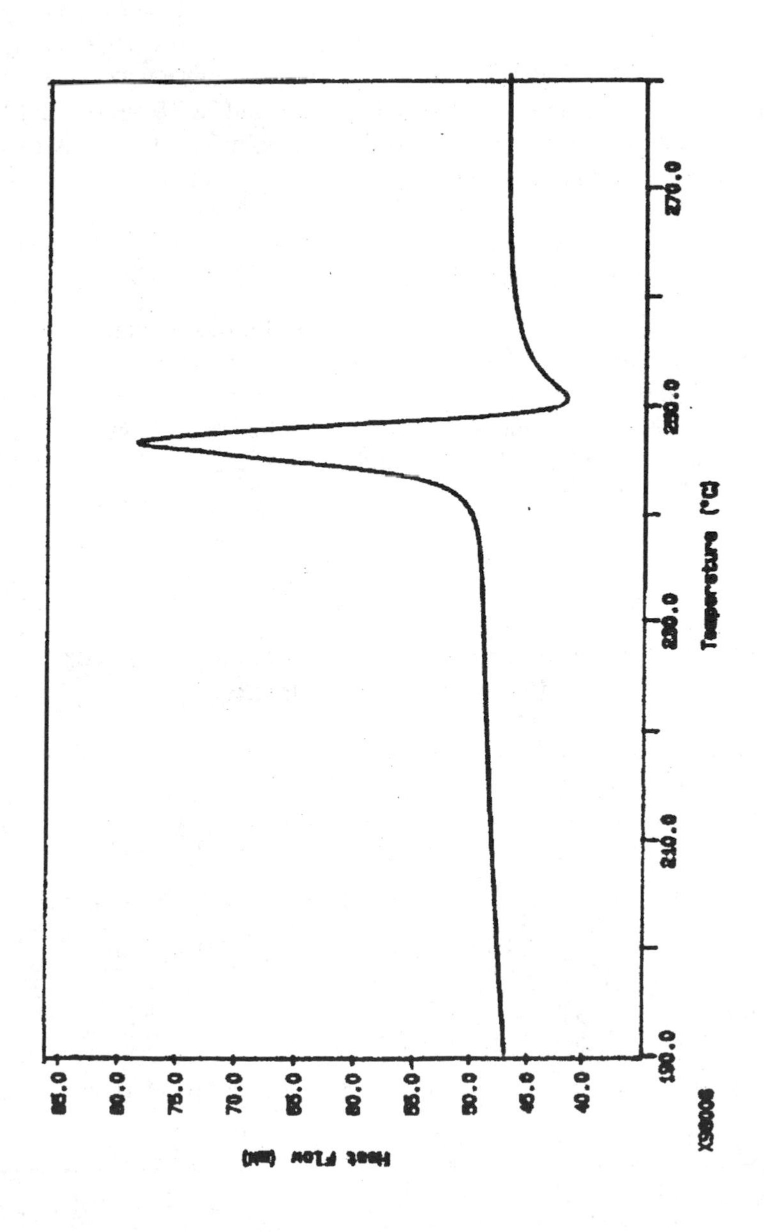

3.6 Spectroscopy

3.6.1 UV/VIS Spectroscopy

The UV absorption spectrum of Flavoxate hydrochloride (dissolved in methanol at a concentration of 11.94 mg/mL) was obtained over a range of 205-400 nm using a Perkin Elmer Lambda 2 spectrophotometer. As illustrated in Figure 3, the compound exhibits absorption maxima at 240.5 nm ($\varepsilon = 1.54 \times 10^4$), 288.2 nm ($\varepsilon = 1.11 \times 10^4$), and 317.8 nm ($\varepsilon = 1.12 \times 10^4$). The UV absorption spectrum of Flavoxate hydrochloride dissolved in 0.1N HCl is similar to the spectrum recorded in methanol [15].

3.6.2 Vibrational Spectroscopy

The infrared spectrum of Flavoxate hydrochloride was obtained on a Perkin Elmer Paragon 1000PC FT-IR spectrometer. The compound was prepared as a potassium bromide pellet, and the resulting IR absorption spectrum is shown in Figure 4. A summary of functional group assignment for the major observed bands is provided in Table 4.

Table 4

Band Assignments for the Infrared Absorption Spectrum of Flavoxate Hydrochloride

Band Energy (cm^{-1})	Assignment
3100-2750	aromatic and aliphatic (C-H) stretch
2444	tertiary amine hydrochloride (N-H) stretch
1718	carboxylic acid ester (C=O) stretch
1636	flavone (C=O) stretch
1616-1440	aromatic (C=C) stretch
1300-1024	carboxylic acid ester and flavone (C-O) stretch
858-648	aromatic (C-H) out-of-plane bend

Figure 3. Ultraviolet absorption spectrum of Flavoxate hydrochloride.

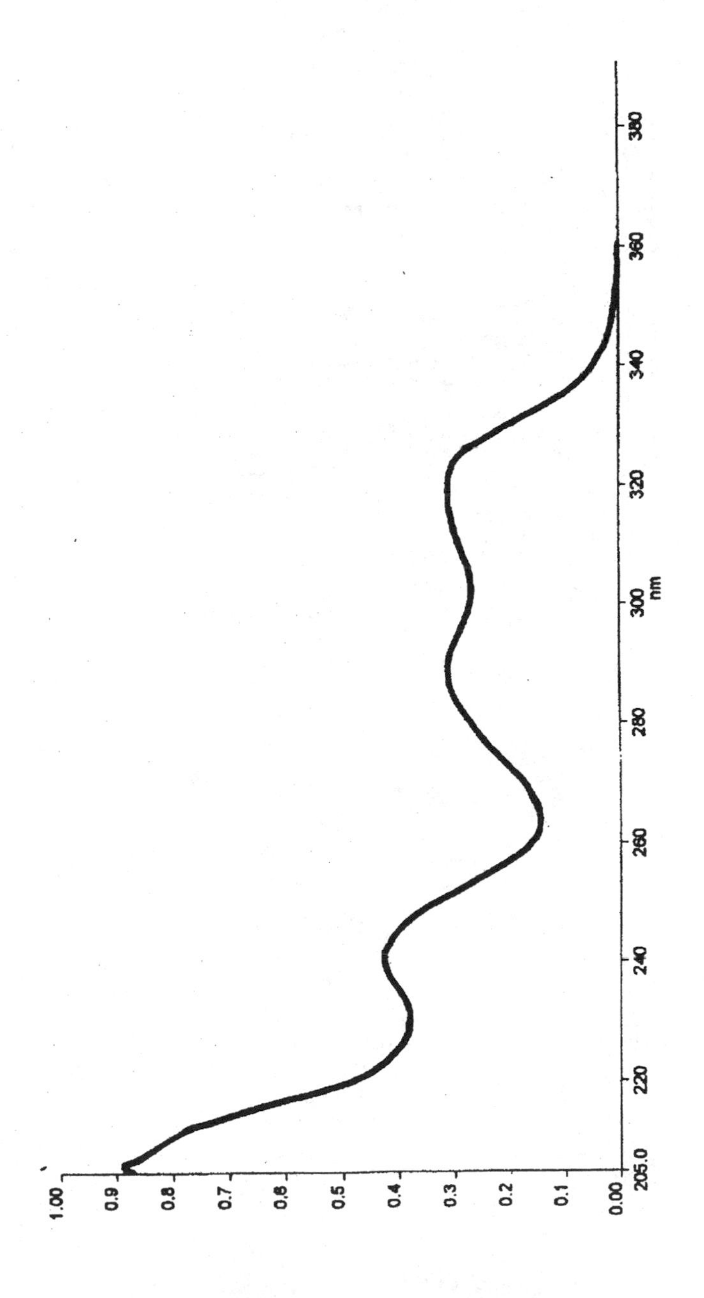

Figure 4. Infrared absorption spectrum of Flavoxate hydrochloride.

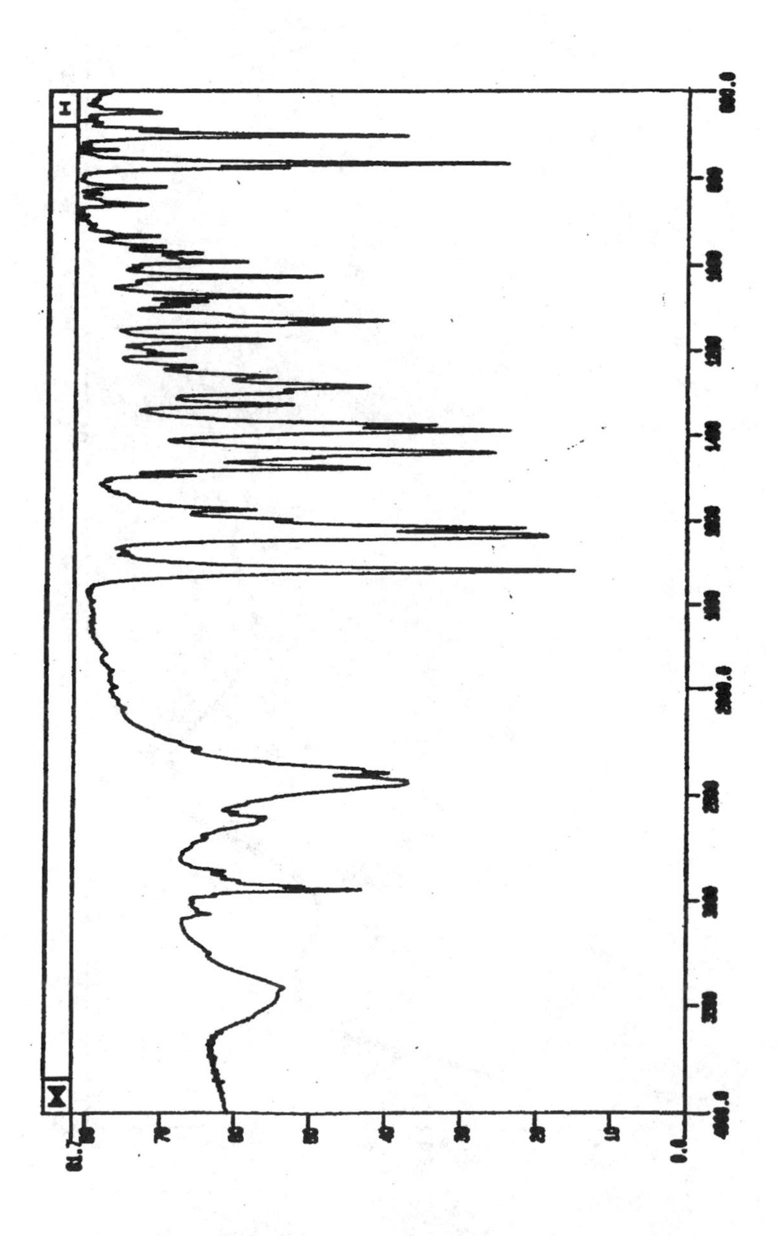

3.6.3 Nuclear Magnetic Resonance Spectrometry

Both the ^{1}H-NMR and ^{13}C-NMR spectra of Flavoxate hydrochloride were obtained on a Bruker ARX-300 spectrometer, operating at 300.157 MHz (^{1}H-NMR) or at 75.478 MHz (^{13}C-NMR). Spectra were recorded in an approximate 1:1 DMSO-d_6 / $CDCl_3$ mixture at room temperature using TMS as the internal standard.

3.6.3.1 ^{1}H-NMR Spectrum

The ^{1}H-NMR spectrum of Flavoxate hydrochloride is shown in Figure 5, and the resonance signal assignments are listed in Table 5.

3.6.3.2 ^{13}C-NMR Spectrum

The ^{13}C-NMR and DEPT ^{13}C-NMR spectra of Flavoxate HCl are shown in Figures 6 and 7, respectively. The resonance signal assignments are provided in Table 6.

3.7 Mass Spectrometry

An electrospray ionization mass spectrometry study of Flavoxate hydrochloride was carried out on a Perkin Elmer/Sciex API-300 triple quadrupole mass spectrometer. The sample was dissolved in aqueous acetonitrile (50/50 v/v) containing 0.05% of acetic acid, and introduced into the mass spectrometer at a flow rate of 0.3 mL/h using a syringe pump. The electrospray ionization mass spectrum (ESI MS) spectrum of Flavoxate hydrochloride is shown in Figure 8. The spectrum displays the protonated Flavoxate molecular ion peak (MH^{+}) at m/z 392. The MS/MS spectrum of this ion is shown in Figure 9. The fragmentation pattern, observed in the MS/MS spectrum, is illustrated in Scheme 2.

4. Methods of Analysis

4.1 Compendial Tests [3]

4.1.1 Identification

4.1.1.1 Infrared Absorption Spectrum

The British Pharmacopoeia (2000 edition) contains the reference IR absorption spectrum of Flavoxate hydrochloride (RS 143). The experimentally determined spectrum, recorded as described in Appendix IIA of the BP 2000, must be concordant with that of the reference spectrum.

Figure 5. ^{1}H-NMR spectrum of Flavoxate hydrochloride.

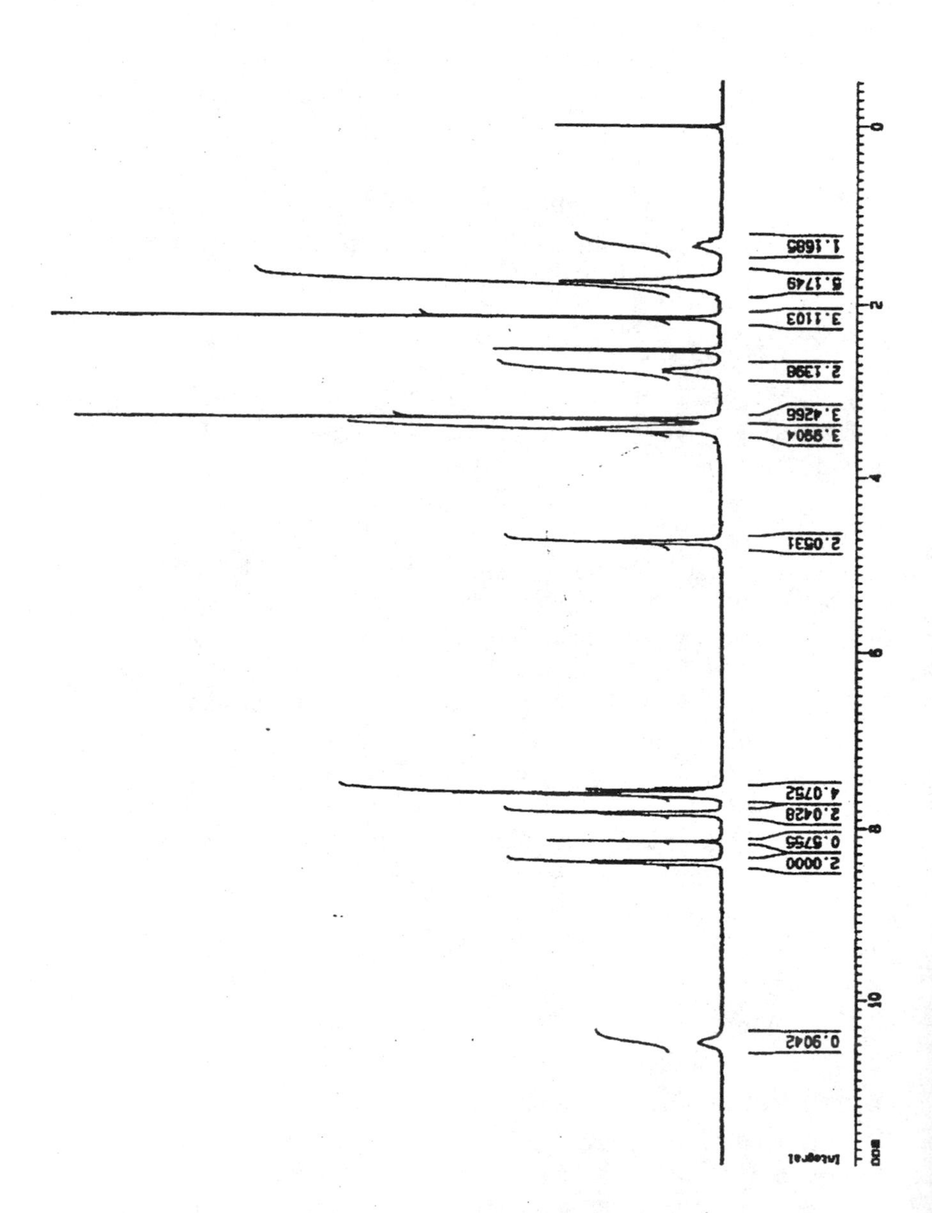

Table 5

^{1}H-NMR Spectral Data for Flavoxate Hydrochloride

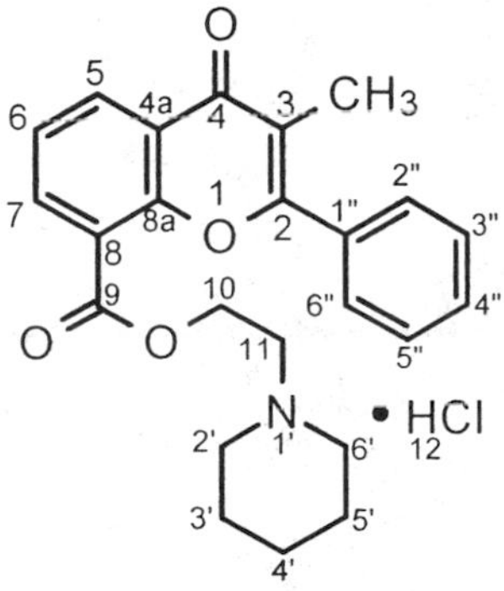

^{1}H Chemical Shift (δ, ppm)	Multiplicity[a]	Integration	Assignment
10.48[b]	bs	1H	H12
8.40	m	2H	H5, H7
7.90-7.78	m	2H	H2", H6"
7.70-7.50	m	4H	H6, H3", H4", H5
4.73	m	2H	H10
3.45	m	4H	H11, H2'a, H6'a[c]
2.78	m	2H	H2'e, H6'e[c]
2.16	s	3H	3-CH_3
1.74	m	5H	H3', H4'e, H5'
1.40-1.23	m	1H	H4'a

[a] bs – broad singlet; m – multiplet; s – singlet
[b] exchangeable with D_2O
[c] a – axial, e - equatorial

Figure 6. ^{13}C-NMR spectrum of Flavoxate hydrochloride.

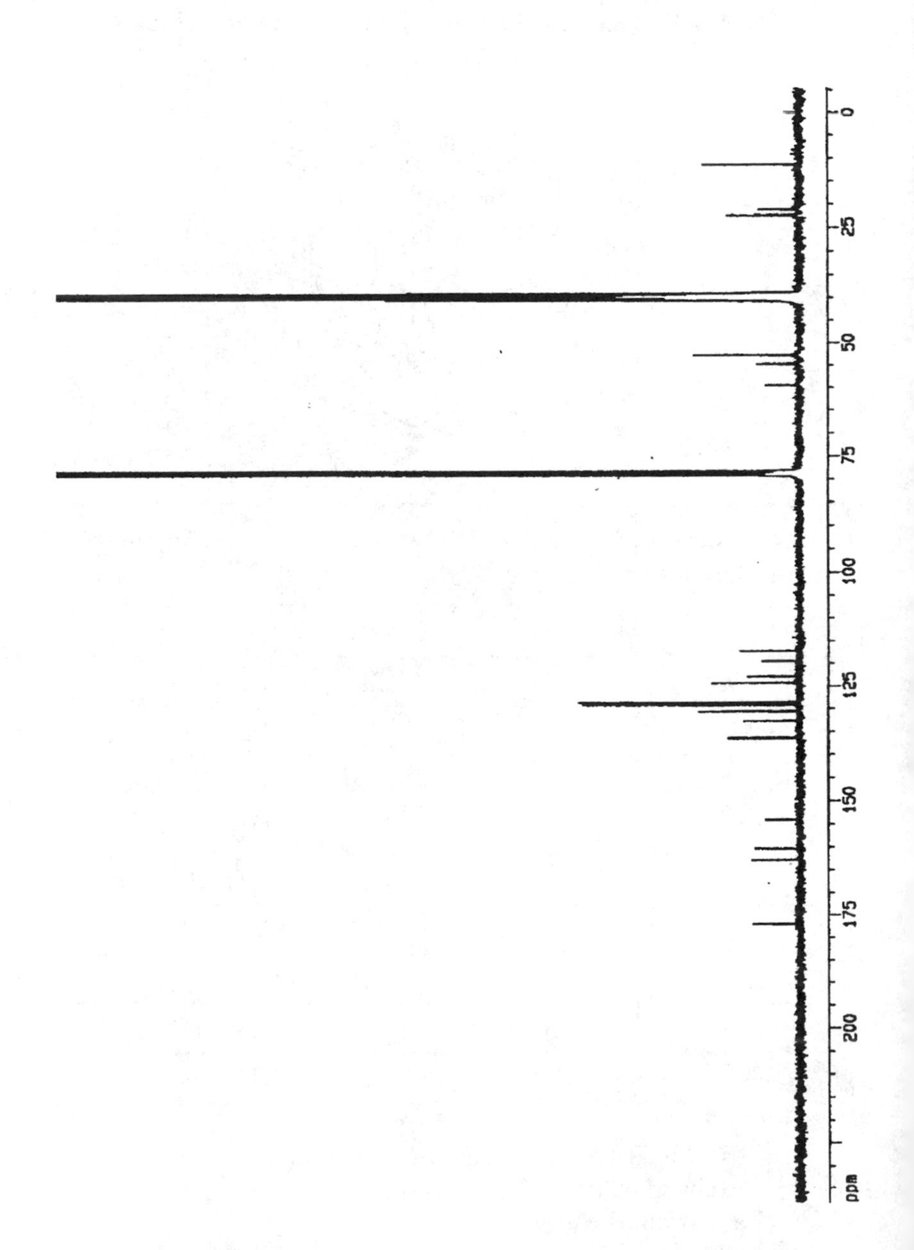

Figure 7. DEPT ^{13}C-NMR spectrum of Flavoxate hydrochloride.

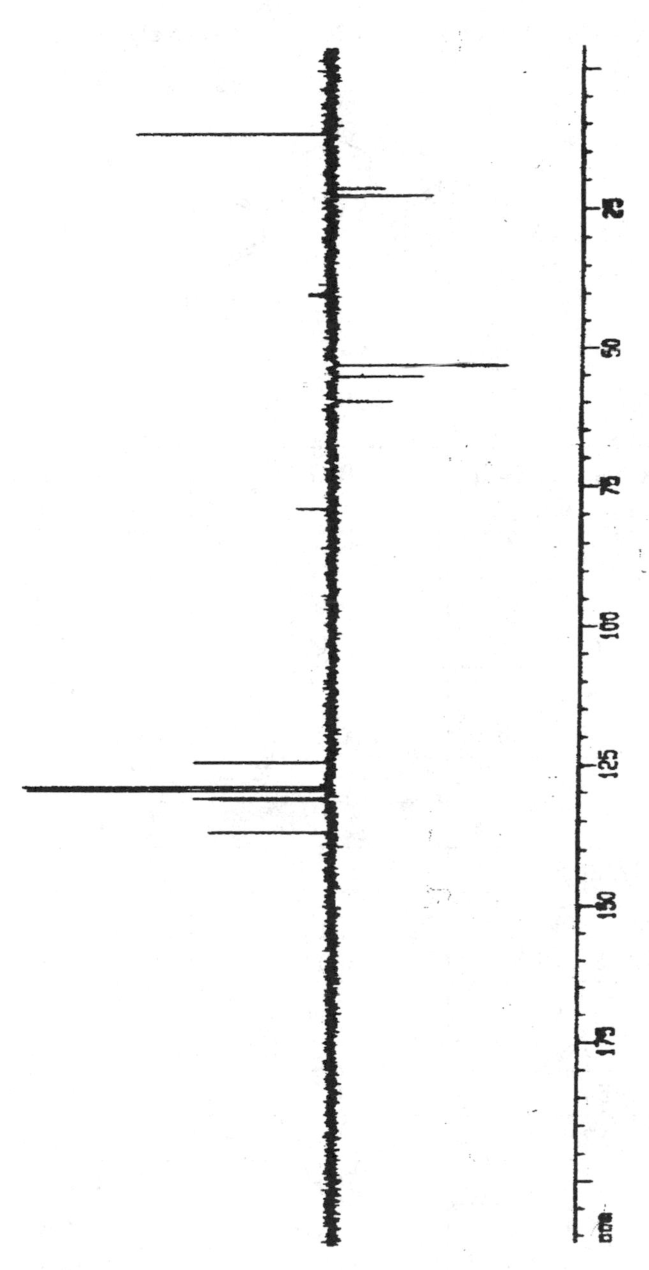

Table 6

^{13}C-NMR Spectral Data for Flavoxate Hydrochloride

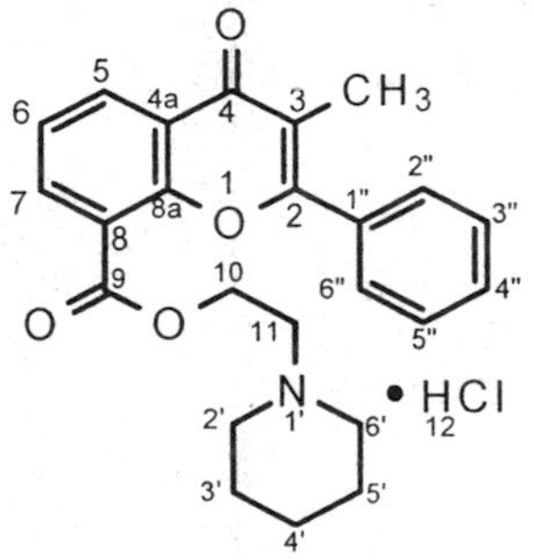

^{13}C Chemical Shift (δ, ppm)	DEPT	Assignment
176.9	C	C4
162.9	C	C9
160.4	C	C2
154.1	C	C8a
136.4, 130.6, 130.5	each CH	C5, C7, C4”
132.6	C	C1”
129.1	2 x CH	C2”, C6”
128.5	2 x CH	C3”, C5”
124.2	CH	C6
122.8, 119.5	both CH	C8, C4a
117.1	C	C3
59.5	CH_2	C10
54.7	CH_2	C11
52.6	CH_2	C2’, C6’
22.4	CH_2	C3’, C5’
21.1	CH_2	C4’
11.4	CH_3	3-CH_3

Figure 8. Electrospray ionization mass spectrum of Flavoxate hydrochloride.

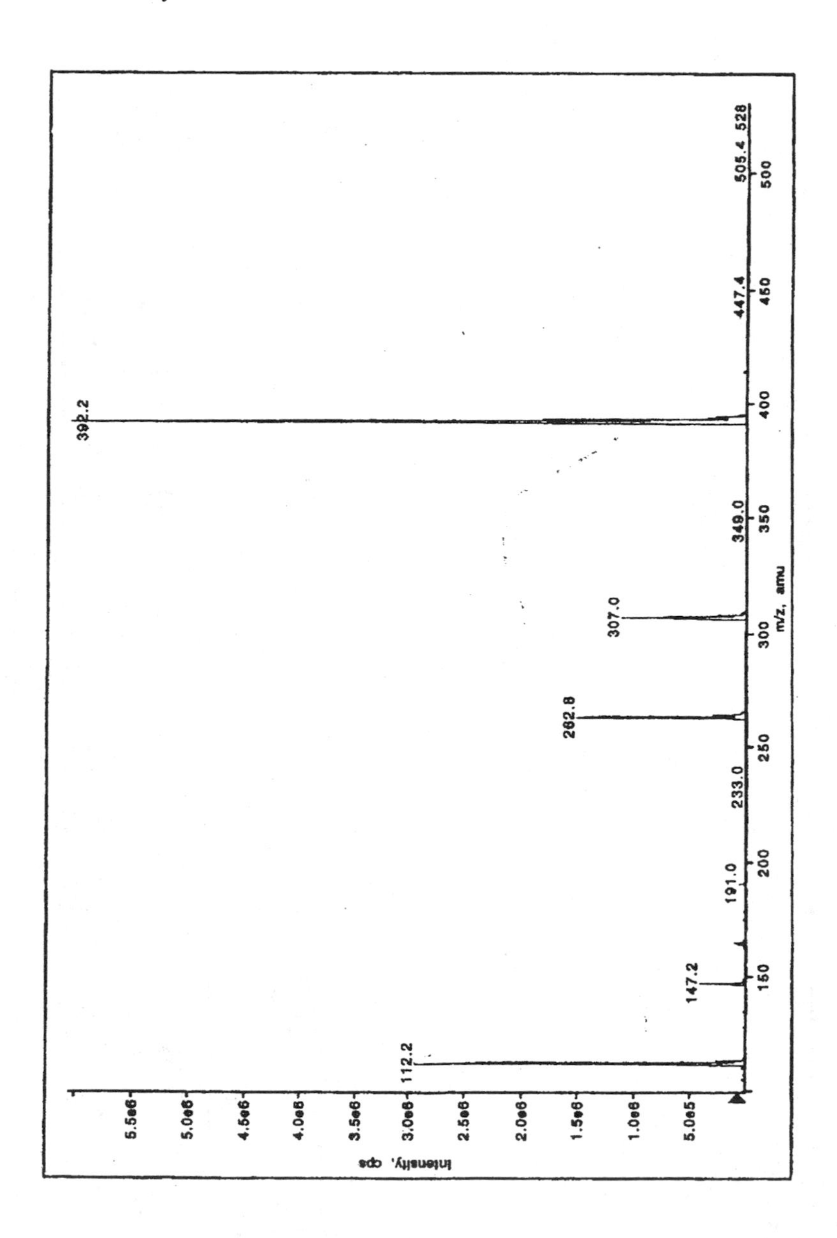

Figure 9. MS/MS spectrum of the protonated Flavoxate molecular ion peak (MH^+) at m/z 392.

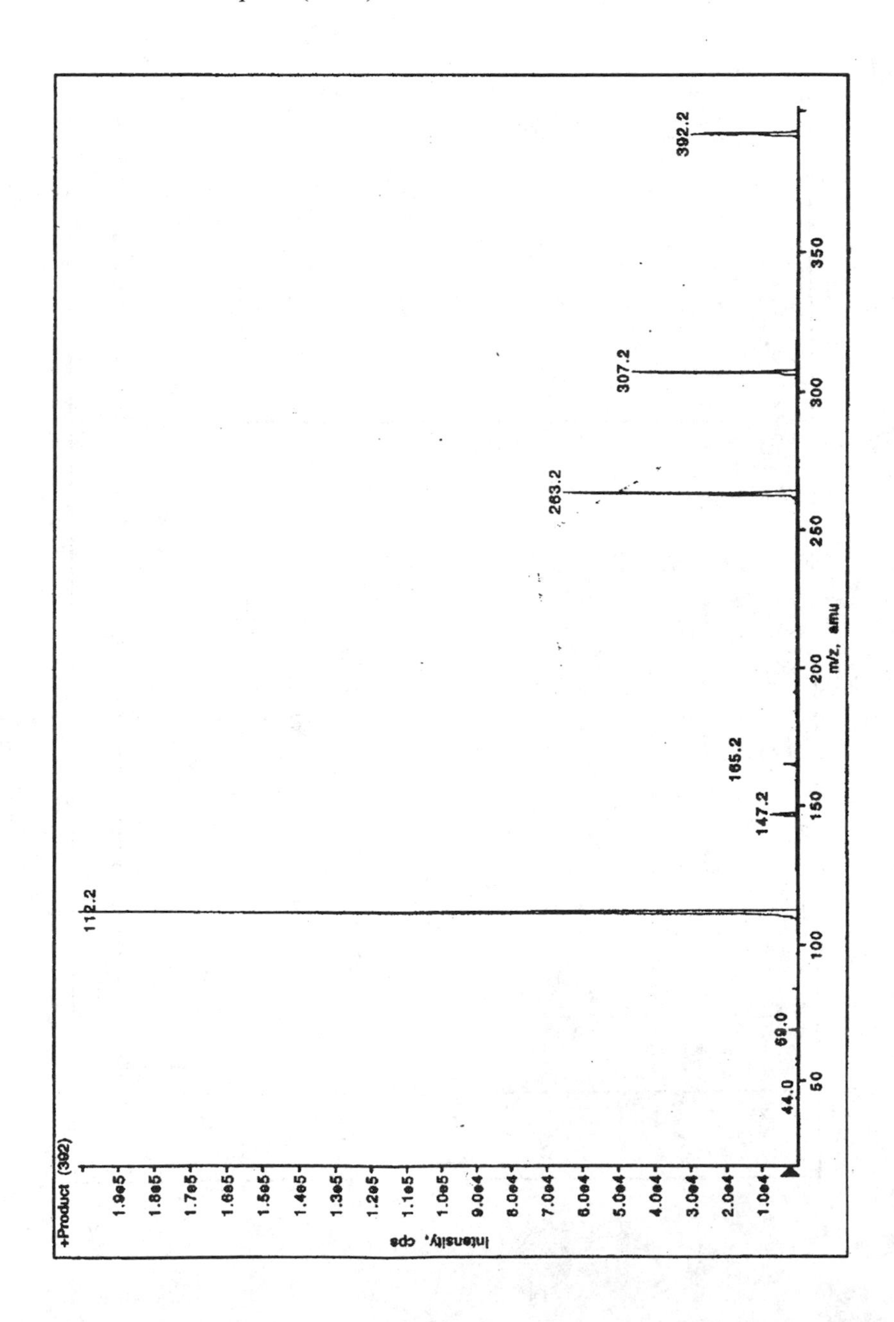

Scheme 2. Fragmentation Pattern from the MS/MS Spectrum of Flavoxate $[MH^{+}]$ Ion.

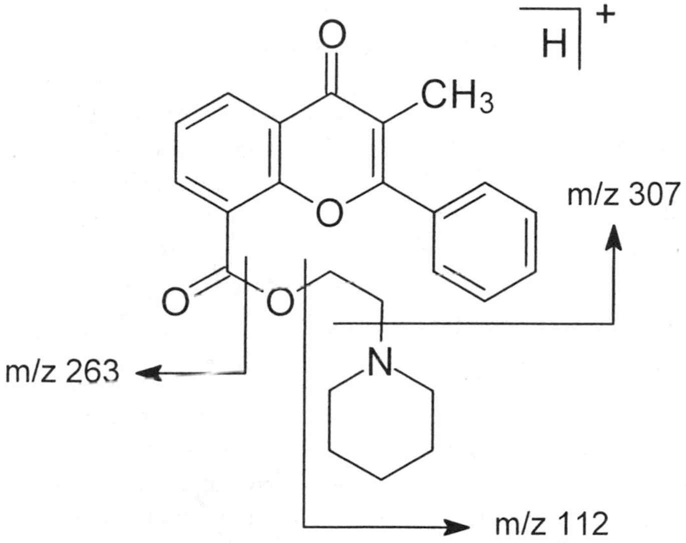

m/z 307

m/z 263

m/z 112

4.1.1.2 Presence of Chloride

The sample must yield reaction A, characteristic of compounds containing chloride. Upon addition of silver nitrate solution to the acidified aqueous solution of the substance, a white precipitate is formed. The confirmatory test entails dissolution of the precipitate in 10M ammonia, as described in Appendix VI of the BP 2000.

4.1.2 Assay

Flavoxate hydrochloride is assayed by non-aqueous titration. The sample (0.5 g) is dissolved in 40 mL of anhydrous acetic acid, and after addition of 10 mL of mercury(II) acetate the solution is titrated with 0.1M perchloric acid using oracet blue B solution as the endpoint indicator. Each milliliter of 0.1M perchloric acid is equivalent to 42.79 mg of Flavoxate hydrochloride. The method was originally suggested and evaluated by Fontani and Setnikar [15].

4.1.3 Related Substances

Two normal-phase thin-layer chromatographic methods are given in the BP 2000 for the determination of related substances. The first method employs a silica gel GF_{254} plate as the stationary phase, and a 1:80:200 mixture (v/v/v) of 18M ammonia, *iso*-propanol and ethyl acetate as the mobile phase. The method is used to determine 3-methylflavone-8-carboxylic acid ethyl ester (a potential synthetic impurity) and unknown impurities. The following four solutions in chloroform are applied separately: (1) 50 μL of a 2.0% w/v solution of the substance being examined, (2) 10 μL of a 0.030% w/v solution of 3-methylflavone-8-carboxylic acid (a potential synthetic impurity/degradation product/metabolite), (3) 10 μL of a 0.015% w/v solution of 3-methylflavone-8-carboxylic acid ethyl ester, and (4) 25 μL of a 0.004% w/v solution of the substance being examined. The plate is examined under UV light (254 nm). The spot corresponding to 3-methylflavone-8-carboxylic acid ethyl ester in chromatogram of (1) is not more intense than the spot in chromatogram of (3) (0.15%), and any other secondary spot in chromatogram of (1) {other than any spot corresponding to 3-methylflavone-8-carboxylic acid from chromatogram of (2)} is not more intense than the spot in chromatogram of (4).

The second method, used to determine 3-methylflavone-8-carboxylic acid, also employs a silica gel G_{254} plate. In this method, the mobile phase consists of glacial acetic acid, ethyl acetate, and cyclohexane (4:25:70 v/v/v). Two chloroform solutions, 50 mL of a 2.0% w/v solution of the substance being examined and 10 mL of a 0.03% w/v solution of 3-methylflavone-8-carboxylic acid, are applied separately. The plate is developed by spraying with dilute potassium iodobismuthate solution. Any spot corresponding to 3-methylflavone-8-carboxylic acid in chromatogram of (1) is not more intense than the spot in chromatogram of (2) (0.3%).

4.1.4 Loss On Drying

The sample (1 g) is dried to constant weight at 105°C. The weight loss is not more than 0.5%.

4.1.5 Heavy Metals

2.0 g complies with limit test C for heavy metals (10 ppm) as per BP 2000, Appendix VII. The standard is prepared using 2 mL of lead standard solution (10 ppm Pb).

4.1.6 Sulfated Ash

When performed according to the procedure given in BP 2000, Appendix IXA, the substance contains not more than 0.1%.

4.2 Titrimetric Analysis

The compendial assay titration method for Flavoxate hydrochloride [3] is described in Section 4.1.2.

4.3 Spectrophotometric Analysis

Two similar ultraviolet spectrophotometric determinations of Flavoxate hydrochloride in its drug product have been reported. In the first method [15], the dosage form containing Flavoxate hydrochloride is dissolved in water, made basic with sodium carbonate, and the liberated base extracted with chloroform. An aliquot of the water-washed chloroform phase is evaporated to dryness, dissolved in 0.1N HCl, and assayed spectrophotometrically at 292 nm. The method is precise and accurate, with average recoveries of 98.5% in ten replicate analyses of Flavoxate hydrochloride tablets.

The second method [17] does not require extraction. Finely ground tablets, equivalent to 100 mg of Flavoxate hydrochloride, are dissolved in 100 mL of 0.1N HCl by heating in a boiling-water bath for 15 minutes. After cooling, the solution is diluted to 250 mL with 0.1N HCl and filtered. A 5 mL portion of the filtrate is further diluted to 100 mL with 0.1N HCl, and the absorbance of this solution measured at 292 nm. The average recovery was 99.9%, and no interference by excipients was noted.

The spectrophotometric determination of Flavoxate hydrochloride and its major metabolite (3-methylflavone-8-carboxylic acid) in plasma will be described in a subsequent section.

4.4 Chromatographic Methods of Analysis

4.4.1 Thin Layer Chromatography

The compendial thin layer chromatographic (TLC) determination of Related Substances in Flavoxate hydrochloride (originally developed and described by Fontani and Setnikar [15]) was described in section 4.1.3.

A TLC methods for the determination of Flavoxate hydrochloride and its main metabolite, 3-methylflavone-8-carboxylic acid, in plasma and urine will be described in a subsequent section.

4.4.2 High Performance Liquid Chromatography

An isocratic reversed phase HPLC method has been developed by the author for the assay of Flavoxate hydrochloride and determination of impurities, and is applicable for both the drug substance and the drug product. The method utilizes an Inertsil ODS-3 column (250 x 4.6 mm, 5 μm) and UV detection at 315 nm. The mobile phase is a 66:32 mixture of 0.02M phosphate buffer (containing 0.01M of tetrabutylammonium hydrogen sulfate and 0.005M of sodium dodecyl sulfate) and tetrahydrofuran; the apparent pH is 5.7; the flow rate is 1 mL/min; the column temperature is 50°C. The method separates potential components within less than 15 minutes run time. A typical chromatogram is shown in

Figure 10. Levels of two potential impurities, 3-methylflavone-8-carboxylic acid and Flavoxate N-oxide, are obtained from the chromatogram, while the level of the UV inactive 1-(2-hydroxyethyl)piperidine (degradation product generated along with 3-methylflavone-8-carboxylic acid) is calculated based on the 1:1 stoichiometry of the hydrolytic cleavage of Flavoxate.

Another isocratic reversed phase HPLC method has been reported for the determination of Flavoxate hydrochloride in pharmaceutical preparations [18]. The method employs a Nucleosil R5 C18 column (125 x 4 mm) and UV detection at 280 nm. The mobile phase is a 1% solution of acetic acid in methanol/water (1:1); the flow rate: 1 mL/min. Diloxanide furoate is used as the internal standard.

Flavoxate hydrochloride can also be analyzed using an isocratic normal phase HPLC procedure that has been developed for 462 basic drugs [19]. The HPLC system comprises a Spherisorb S5W silica column (125 x 4.9 mm, 5 μm), UV (254 nm) and electrochemical detection (+1.2 V). The mobile phase is methanolic ammonium perchlorate (10 mM, pH 6.7).

The HPLC methods used for determination of Flavoxate hydrochloride and its metabolites in biological fluids will be described in a subsequent section.

4.5 Determination in Body Fluids and Tissues

4.5.1 Spectrophotometric Methods

UV spectrophotometry has been used for the assay in plasma of Flavoxate hydrochloride and its main metabolite, 3-methylflavone-8-carboxylic acid [16]. The method is based upon extraction of these compounds with chloroform from the acidified aqueous material, and subsequent UV measurement at 319 nm where both substances have the same molecular absorbance. The method therefore gives the total amount in the sample of compounds having a flavone structure. The UV absorbance is compared with that of the standard solution of Flavoxate hydrochloride in chloroform. The recovery from plasma samples is 95-100%, and the limit of quantitation is 2 μg/mL for Flavoxate hydrochloride or its metabolite.

Figure 10. HPLC chromatogram of Flavoxate hydrochloride in the presence of potential impurities:
FW RC1 = 3-methylflavone-8-carboxylic acid
FW RC2 = Flavoxate N-oxide

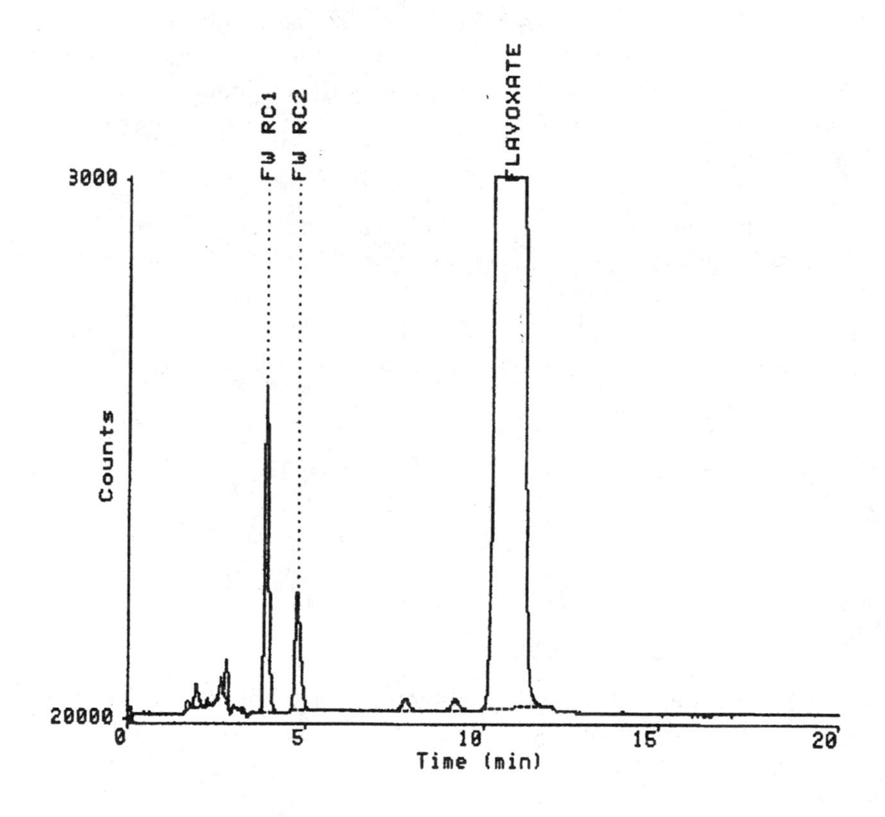

4.5.2 Thin-Layer Chromatography

Thin layer chromatography has been used (Method I) for the assay of 3-methylflavone-8-carboxylic acid in urine after acidic hydrolysis of the sample, where Flavoxate hydrochloride and its metabolite are not distinguished. In Method II a determination is made of the ratio between these two compounds [16]. In method I, the sample of urine is hydrolyzed with concentrated HCl at about 100°C for 2.5 hours to assure a complete conversion of Flavoxate hydrochloride to 3-methylflavone-8-carboxylic acid. The sample is extracted with chloroform, evaporated to dryness, re-dissolved in ethanol, and applied to a Merck Kieselgel 60 F_{254} plate. The chromatogram is run with cyclohexane : ethyl acetate : acetic acid (70:100:4). Quantification is performed on a spectrodensitometer at 320 nm, and is based upon comparison with separately applied standard solutions of 3-methylflavone-8-carboxylic acid.

In Method II [16], the sample of plasma or urine, acidified with 5N HCl, is extracted with chloroform, evaporated to dryness, re-dissolved in chloroform, and chromatographed first as described for Method I to a migration of 15 cm. After drying the plate, a second run for a 5 cm migration is performed using isopropyl alcocol : ethyl acetate : ammonia (40:100:0.5). Standards of Flavoxate hydrochloride and 3-methylflavone-8-carboxylic acid are applied separately on the same plate. The ratio between these two compounds is obtained using a spectrodensitometer as in Method I.

4.5.3 High Performance Liquid Chromatography

HPLC systems for determination of Flavoxate hydrochloride and its main metabolite in biological fluids are described in Table 7.

4.5.4 Gas Chromatography

Similar to the TLC method previously described, gas chromatography has been used for determination of 3-methylflavone-8-carboxylic acid in urine after an acidic hydrolysis [16]. This technique does not distinguish between Flavoxate hydrochloride and its main metabolite [16]. The method utilizes the chloroform extract obtained as described for TLC Method I, and is based upon derivatization of the metabolite (3-methylflavone-8-carboxylic acid) to the corresponding methyl ester using

Table 7

HPLC Systems for Determination of Flavoxate Hydrochloride and 3-Methylflavone-8-Carboxylic Acid in Biological Fluids

Column	Temp., °C	Mobile Phase	Detection	Application	Reference
Cosmosil $5C_{18}$ (150 x 4.6 mm)	Ambient	0.2M H_3PO_4-KH_2PO_4 buffer (pH 3.0) : acetonitrile: ethanol: water (3:10:2:11) Flow rate: 1 mL/min	UV (318 nm)	Determination of flavoxate and its main metabolite in rat plasma, urine and bile	20
YWG ODS (250 x 4.6 mm, 10 μ)	Ambient	Methanol : 20 mM KH_2PO_4 (pH 2.5) buffer (70:30) Flow rate: 1.3 mL/min	UV (310 nm)	Determination of the main metabolite in human plasma	21
Nova-pak C8 (150 x 3.9 mm, 5 μ)	40°C	Acetonitrile : 25 mM KH_2PO_4 (pH 2.5) buffer Flow rate: 1 mL/min	UV (319 nm)	Determination of the main metabolite in human plasma	This work

diazomethane. The GC system employs a coiled glass column (3 ft x 1/8 in) and flame ionization detector. The stationary phase is 0.8% OV-1 on re-silanized 100-120 mesh Gas Chrom Z. Helium (30 mL/min) is used the carrier gas, and the column temperature is 218°C. 3-Methylflavone-8-carboxylic acid ethyl ester is used as the internal standard.

In another method, 3-methylflavone-8-carboxylic acid was determined in human plasma and urine after oral administration of Flavoxate chloride [22]. The samples were hydrolyzed, and then the metabolite extracted with benzene and derivatized by methylation with methanol in diluted aqueous HCl. The GC analysis was carried out on a U-shaped glass column (1.2 m x 0.3 cm) containing 4% SE-30 on Chromosorb G (60-80 mesh) at 260°C using helium (80 mL/min) as the carrier gas. The method also employed 7-bromo-1,3-dihydro-1-methyl-5-(2-fluorophenyl)-2H-1,4-benzodiazepine-2-on as the internal standard. This method was also suitable for GC-MS analysis. The GC conditions were the same, except that a column containing 1% OV-17 was utilized [22].

4.5.5 Capillary Electrophoresis

Capillary electrophoresis has been used to determine 3-methylflavone-8-carboxylic acid in human urine [23]. The system consisted of a BioRad uncoated fused-silica capillary (35 cm x 50 μm), UV detection at 245 nm, 0.02M phosphate buffer (pH 7), containing 0.01M of tetramethylammonium bromide and 0.02M of sodium dodecyl sulfate, and applied voltage of 10 kV. Electrochemical injections were performed at 2 kV for 10 seconds.

5. Stability

5.1 Solid-State Stability

Flavoxate hydrochloride has been found to be a highly stable compound in the solid state. No signs of degradation were noted in dry material stored for more than three years at room temperature [15]. The stress study performed in this laboratory showed that Flavoxate hydrochloride remains intact when exposed to elevated temperature (60-80°C, 4 weeks), UV/visible light (380-770 nm, 650 W/m^2, 16 hours), or when stored in a high humidity chamber (40°C/75% relative humidity, 4 weeks).

The drug product (Flavoxate hydrochloride 200 mg tablets) is also stable. Even under harsh thermal stress conditions (4 week storage at 80°C), it undergoes very insignificant degradation resulting in hydrolytic cleavage of the ester moiety to give low levels (0.7-1.5%) of the corresponding degradants (see Scheme 3). Not more than 0.1% of known and unknown impurities were detected in light and high humidity stressed tablets. Flavoxate hydrochloride tablets should be stored at 15-30°C, and have an expiration date of 3 years following the date of manufacture [24].

5.2 Solution-Phase Stability

In aqueous media, the ester function of Flavoxate hydrochloride undergoes hydrolysis at a rate which is pH dependent [15,16]. When a 0.5% solution of Flavoxate hydrochloride in 0.1N HCl was stored for 30 days at 37°C, the degradation was 9.3%, while at reflux temperature the degradation was 10% and 30% after 3 and 8 hours, respectively. The degradation profile in water is similar owing to the presence of HCl as part of the drug substance itself. In alkaline medium the hydrolysis occurs faster. When Flavoxate hydrochloride is dissolved in a 1:1 acetone/pH 10 phosphate buffer, it loses 82% of its potency after 24 hours at 20°C [15]. In a pH 7.4 phosphate buffer, Flavoxate hydrochloride was 50% hydrolyzed after 60 minutes. In pH 7.4 phosphate buffer containing 0.2% bovine serum albumin, the half-hydrolysis time was 30 minutes [16].

A stress study carried out in this laboratory confirmed the susceptibility of Flavoxate hydrochloride to hydrolysis in aqueous media. The degradation of Flavoxate hydrochloride in 0.01N HCl (reflux temperature, 4 hours), phosphate buffer pH 9.3 (80°C, 30 minutes) and water (reflux temperature, 4 hours) was about 11%, 21% and 21%, respectively. The hydrolysis results in formation of 3-methylflavone-8-carboxylic acid and 1-(2-hydroxyethyl)piperidine, as shown in Scheme 3. In 3% aqueous hydrogen peroxide solution, Flavoxate hydrochloride undergoes slow oxidation to give the corresponding N-oxide. The level of this substance was 0.5% after 3 hour reaction at room temperature. Irradiation of aqueous solution of Flavoxate hydrochloride with UV/visible light at room temperature results in insignificant hydrolytic degradation.

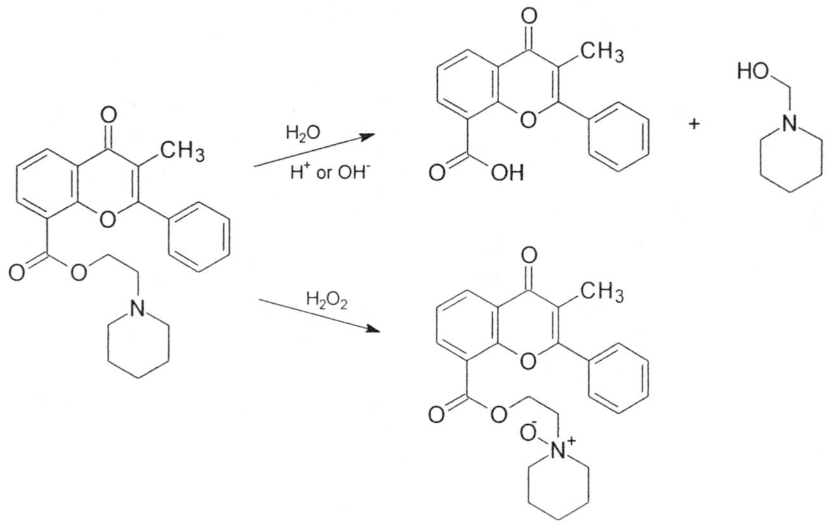

Scheme 3 Degradation Pathways of Flavoxate in Aqueous Solution.

5.3 Stability in Biological Fluids

Flavoxate hydrochloride is not very stable in biological samples. The following half-hydrolysis times were found for 100 μg/mL solutions of the drug in plasma [16]:

Species	20°C	37°C
Rat	120	15
Rabbit	5	5
Dog	1260	30
Man	120	60

6. Drug Metabolism and Pharmacokinetics

6.1 Adsorption and Distribution

The pharmacokinetics study in man revealed that after oral administration, Flavoxate hydrochloride is completely and rapidly adsorbed from the gastrointestinal tract [5-7, 25, 26]. The AUC after oral administration is similar to that obtained after intravenous administration. A comparative study in male man subjects showed that the ratio between the quantities excreted after administration of the sugar coated tablets and of the solution (*i.e.*, the index of bioavailability) was 1.1 [25]. The absolute bioavailability of Flavoxate hydrochloride is 90% [27]. Distribution was determined in rats and dogs following oral and intravenous administration of C^{14}-labeled Flavoxate. Tissue distribution was low in the brain, but high in the liver, kidney and bladder [6].

6.2 Metabolism and Elimination

The pharmacokinetics study in man showed that after intravenous administration the drug disappears from blood with a half-life of about 5 minutes, and after 1 hour the plasma levels became very low [25]. After administration, Flavoxate hydrochloride is converted almost immediately to 3-methylflavone-8-carboxylic acid (parallel formation of 1-(2-hydroxyethanol)piperidine must also occur). The acid accounts for 40-50% of the total amount of urinary metabolites maintaining the chromone ring. The remaining 50-60% can be recovered also as 3-methylflavone-8-carboxylic acid upon acidic or glucuronidasic hydrolysis [25]. Flavoxate as such is not present in urine even after intravenous administration. After oral administration of 200 mg and 400 mg doses to man, small concentrations of Flavoxate hydrochloride (0.3-0.7 μg/mL) were detected in plasma 1 hour after dosing. Concentrations of the major metabolite was 5-10 μg/mL in 1 hour, increasing to 9-15 μg/mL within 2 hours [28]. After intravenous or oral administration, about half of the administered Flavoxate hydrochloride is excreted in urine (as 3-methylflavone-8-carboxylic acid). The other half is likely to be excreted through the bile as in the rat [29].

According to the C^{14}-labeled Flavoxate study on rats and dogs, urinary and fecal excretion was complete in 24 hours. About 30-40% of the C^{14} appeared in the urine and 50-60% concentrated in the feces [6].

7. Toxicity

The intravenous LD_{50} in rats was determined to be 27.8 mg/kg [11]. As shown in the following table [28], Flavoxate hydrochloride was of low toxicity when fed to experimental animals.

Species	LD_{50} (g/kg)
Rat	4.75/1.24[a]
Rabbit	3
Dog	2
Mouse	1.36

[a]Adult / Weanling

No significant toxicological findings were made in chronic toxicity studies performed on rats and dogs [28].

Acknowledgments

The author would like to thank a number of people from Apotex R&D who have contributed to the present review. I am indebted to Dr. Nicholas Cappuccino and Ms. Elisabeth Kovacs for encouragement, management support, and valuable discussions. I am grateful to Dr. Tom Hu and Mr. Michael Tang for performing physical testing and to Dr. Maria Foster for her assistance in interpreting spectroscopic data. Thanks are due to Mr. Dragisa Jovanovic who developed the HPLC methods for Flavoxate hydrochloride drug substance and drug product and who performed the stress study. I should also like to express my gratitude to Ms. Sandy Du, Dr. Bang Qian Xu, Mr. Doug Watson, and Mr. Paul Gordon for providing information regarding the HPLC method for determination of Flavoxate metabolite in human plasma. The help of Ms. Karen Taylor, Ms. Annette Goldberger, and Ms. Tze-Ling Kong in collecting the literature pertinent to the present subject is gratefully acknowledged.

References

1. ***The Merck Index***, 12th Edition, Merck & Co., Inc., Whitehouse Station, N.J., 1996, p. 693.

2. ***USP Dictionary of USAN and International Drug Names***, The United States Pharmacopoeial Convention, Inc., Rockville, MD, 2000, p. 305.

3. ***British Pharmacopoeia***, Vol. I, The Stationery Office, London, 2000, p. 690.

4. I. Setnikar, M.T. Ravasi, and P. Da Re, *J. Pharm. Exp. Therap.*, **130**, 356 (1960).

5. L. Guarnery, E. Robinson, and B. Testa, *Drugs Today*, **30**, 91 (1994).

6. ***Compendium of Pharmaceuticals and Specialties***, 35th Edition, Canadian Pharmacists Association, Ottawa, 2000, p. 1660.

7. R. Ruffmann and A. Sartani, *Drugs Exptl. Clin. Res.*, **13**, 57 (1987)

8. S. Gould, *Urology*, **5**, 612 (1975).

9. A. Zanollo and F. Catanzaro, *Urol. Int.*, **35**, 176 (1980).

10. D.V. Bradly and R.J. Cazort, *J. Clin. Pharm.*, **10**, 65 (1970).

11. C. Pietra and P. Cazzulani, *Il Pharmaco. Ed. Pr.*, **41**, 267 (1986).

12. P. Cazzulani, R. Panzarasa, C. De Stefani, and G. Graziani, *Arch. Int. Pharmacodyn.*, **274**, 189 (1985).

13. P. Da Re, L. Verlicchi, and I. Setnikar, *J. Med. Pharm. Chem.*, **2**, 263 (1960).

14. P. Da Re, *Il Pharmaco. Ed. Sci.*, **11**, 662 (1956).

15. F. Fontani, I. Setnikar, *Pharm. Ind.*, **36**, 802 (1974).

16. A. Cova, I. Setnikar, *Arzeim.-Forsch. (Drug Res.)*, **25**, 1707 (1975).

17. Y. Zheng, *Yaowu Fenxi Zazhi*, **13**, 339 (1993).

18. S.S. Zarapkar, U.B. Salunkhe, B.B. Sanunkhe, V.J. Doshi, S.V. Sawant, and R.V. Rele. *Indian Drugs*, **26**, 354 (1989).

19. I. Jane, A. McKinnon, and R.J.Flanagan. *J. Chrom.*, **323**, 191 (1985).

20. Y. Kaneo, T. Morochika, T. Tanaka, K. Ozawa, S. Kojima, K. Suguro, and K. Abiko, *Iyakuhin Kenkyu*, **21**, 694 (1990).

21. Y. Huang, Q. Yu, M.-Z. Liang, J.-X. Wang, Y.-P. Qin, and Y-G. Zou, *West China J. Pharm. Sci.*, **15**, 9 (2000).

22. T. Ariga, K. Tanaka, K. Hattori, M. Hioki, H. Shindo, *Ann. Sankyo Res. Lab.*, **26**, 94 (1974).

23. C.-X. Zhang, Z.-P. Sun, D.-K. Ling, J.-S. Zheng, J. Guo, and X.-Y. Li, *J. Chrom.*, **612**, 287 (1993).

24. ***AHFS Drug Information***, G.K. McEvoy, ed., The American Society of Health-System Pharmacists, Maryland, 2000, p. 3292.

25. M. Bertoli, F. Conti, M. Conti, A. Cova, and I. Setnikar, *Pharm. Res. Commun.*, **8**, 417 (1976).

26. R. Ruffmann, *J. Int. Med. Res.*, **16**, 317 (1988).

27. W.K. Sietsema, *Int. J. Clin. Pharmacol.*, **27**, 179 (1989).

28. ***URISPAS (Brand of Flavoxate Hydrochloride) Product Monograph***, Paladin Labs Inc., Montreal, 1988.

29. I. Setnikar, A. Cova, M.J. Magistretti, *Arzniem.-Forsh. (Drug Res.)*, **25**, 1916 (1975).

LANSOPRAZOLE

J. Al-Zehouri, H.I. El-Subbagh, and Abdullah A. Al-Badr

Department of Pharmaceutical Chemistry,
College of Pharmacy
King Saud University,
P.O. Box 2457, Riyadh-11451
Saudi Arabia

1075-6280/01 $35.00

Contents

1. Description

1.1 Nomenclature

1.1.1 Systematic Chemical Names [1-3]

2-(2-Benzimidazolylsulfinylmethyl)-3-methyl-4-(2,2,2-trifluro-ethoxy)pyridine.

2-[[[3-Methyl-4-(2,2,2,-trifluoroethoxy)-2-pyridinyl]methyl]-sulfinyl]-1H benzimidazole.

2-[4-(2,2,2-Trifluoroethoxy)-3-methyl-2-pyridinylmethylsulfinyl] 1H benzimidazole.

3-{[3-Methyl-4-(2,2,2-trifluoroethoxy)-2-pyridyl]methyl}-sulfinyl benzimidazole.

1.1.2 Nonproprietary Name

Lansoprazole

1.1.3 Proprietary Names [1-3]

Agopton, Bamalite, Dakar, Lansox, Lanzor, Limpidex, Ogast, Opiren, Prevacid, Prezal, Takepron, Zoton

1.2 Formulae

1.2.1 Empirical Formula, Molecular Weight, CAS Number

$C_{16}H_{14}F_3N_3O_2S$ [MW = 369.363]

CAS number = 103577-45-3

1.2.2 Structural Formula

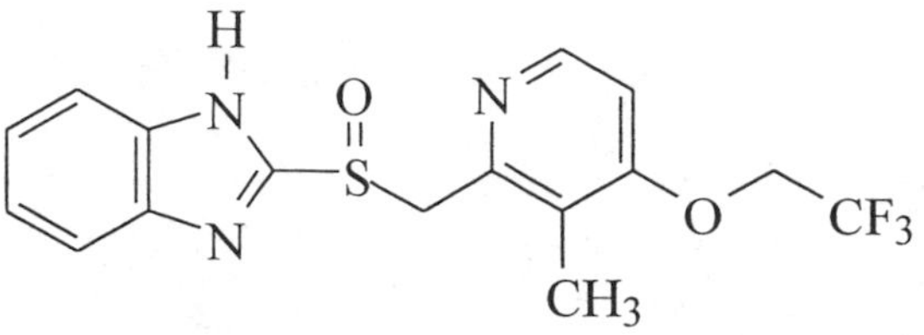

1.3 Elemental Analysis [1]

The calculated elemental composition of Lansoprazole is as follows:

carbon:	52.03%
hydrogen:	3.82%
oxygen:	8.66%
nitrogen:	11.38%
sulfur:	8.66%
fluorine	15.43%

1.4 Appearance

Lansoprazole is a white crystalline powder that is odorless and tasteless.

1.5 Uses and Applications

Lansoprazole is a new, highly potent proton pump (H^+, K^+-ATPase) inhibitor with potent anti-secretory effects. It was demonstrated to be effective in the treatment of duodenal and gastric ulcers, reflux esophagitis, and Zollinger-Ellison syndrome. The compound is rapidly absorbed after oral administration, with peak plasma concentrations being achieved within approximately 1.5 hours. Lansoprazole is substantially metabolized by the liver [3-7].

2. Method(s) of Preparation

Lansoprazole was synthesized and developed in Japan [8], and the synthetic route for its preparation is outlined in Scheme 1. The N-oxide of 2,3-dimethylpyridine (**1**) was nitrated to give the 4-nitro derivative (**2**). The ring is sufficiently activated by the N-oxide group so that treatment with the anion of trifluoroethanol results in the replacement of the nitro group to give the ether (**3**), probably by an addition elimination sequence. Treatment of compound (**3**) by acetic anhydride leads first to the formation of transient o-acetate; this undergoes migration to the adjacent ring methyl group with simultaneous oxidation (Polonovski reaction) to afford (**4**). Saponification of the acetate initially obtained gives the hydroxymethyl derivative (**5**), and then reaction with thionyl chloride converts the hydroxyl group to the chloro (**6**). Alkylation of compound (**6**) with 2-mercaptobenzimidazole is proceeded by an attack by the nucleophilic sulfur. Oxidation of the alkylation product (**7**) with a controlled amount of *m*-chloroperbenzoic acid (*m*-CPBA) yielded Lansoprazole (**8**) [9,10].

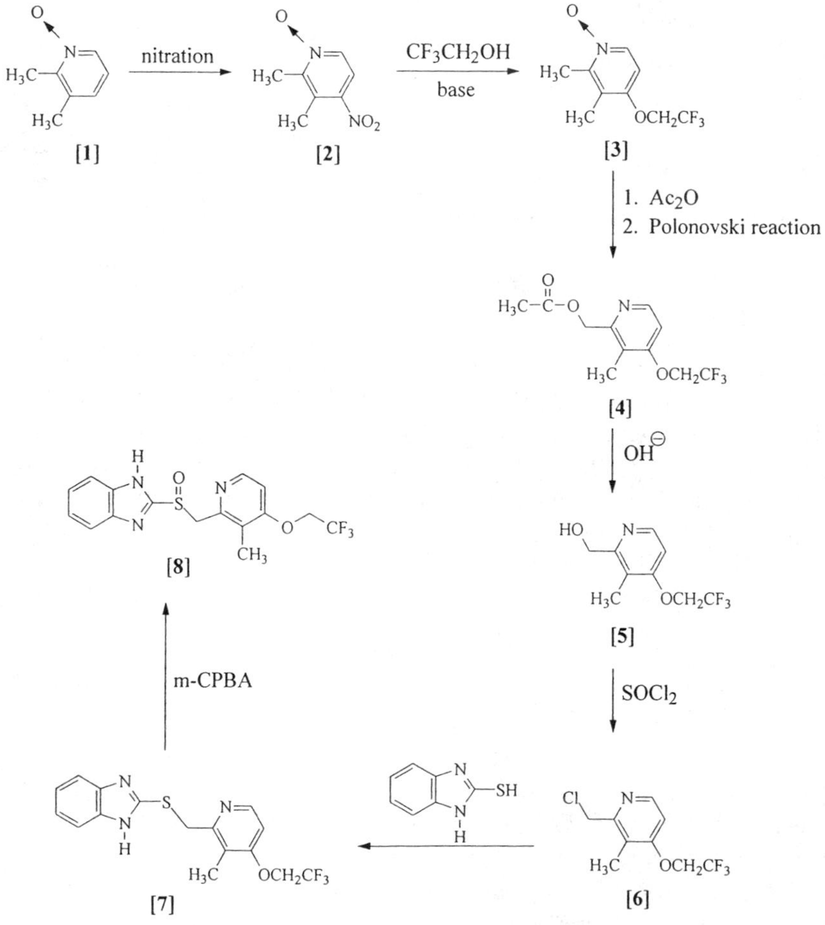

Scheme 1. Synthesis of Lansoprazole.

3. Physical Properties

3.1 Ionization Constants

Using the predictive ACD PhysChem program (Advanced Chemistry Development, Toronto, CA), the following ionization constants were calculated:

$pK_1 = 2.34 \pm 0.37$

$pK_2 = 3.53 \pm 0.37$

$pK_3 = 8.48 \pm 0.30$

3.2 Solubility Characteristics

Lansoprazole has been reported to be soluble in methanol, chloroform, but only very slightly soluble in water [1].

Using ACD PhysChem, and assuming the melting point of Lansoprazole to be 181°C, the following pH dependence of water solubility was calculated:

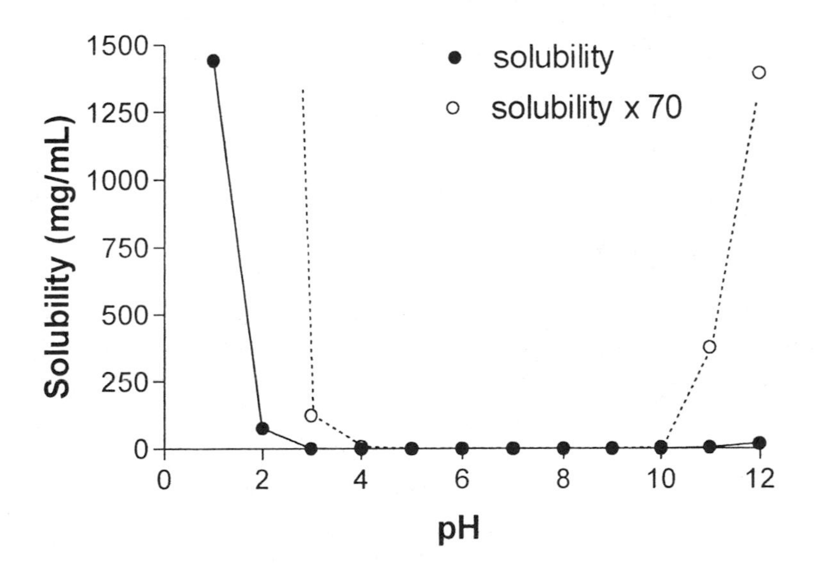

3.3 Partition Coefficient

Using ACD PhysChem, the octanol/water partition coefficient of neutral Lansoprazole was calculated as $\log P = 2.39 \pm 0.78$.

Assuming the melting point of Lansoprazole to be 181°C, ACD PhysChem yielded the following values for the pH dependence of the distribution coefficient:

pH	log D	pH	log D
1.0	–1.20	7.0	2.38
2.0	0.44	8.0	2.27
3.0	1.57	9.0	1.76
4.0	2.21	10.0	0.87
5.0	2.37	11.0	–0.038
6.0	2.39	12.0	–0.60

3.4 X-Ray Powder Diffraction

X-ray powder diffraction was conducted using a Philips PW-1710 diffractometer, equipped with a single crystal monochromator, and using the Kα line of copper as the radiation source. The powder pattern is shown in Figure 1, and the crystallographic results are found in Table 1.

3.5 Thermal Methods of Analysis

3.5.1 Melting Behavior

Lansoprazole is observed to melt within the range of 178-182°C, but with decomposition [1].

3.5.2 Differential Scanning Calorimetry

Differential scanning calorimetry (DSC) of Lansoprazole was performed using a A Dupont TA-9900 thermal analyzer attached to Dupont Data unit. The data were obtained at a scan rate of 10°C/min, over the range of 40-400°C. The thermogram is shown in Figure 2, where it may be observed that the compound melts at 180.95°C.

Figure 1: X-ray powder diffraction pattern of Lansoprazole.

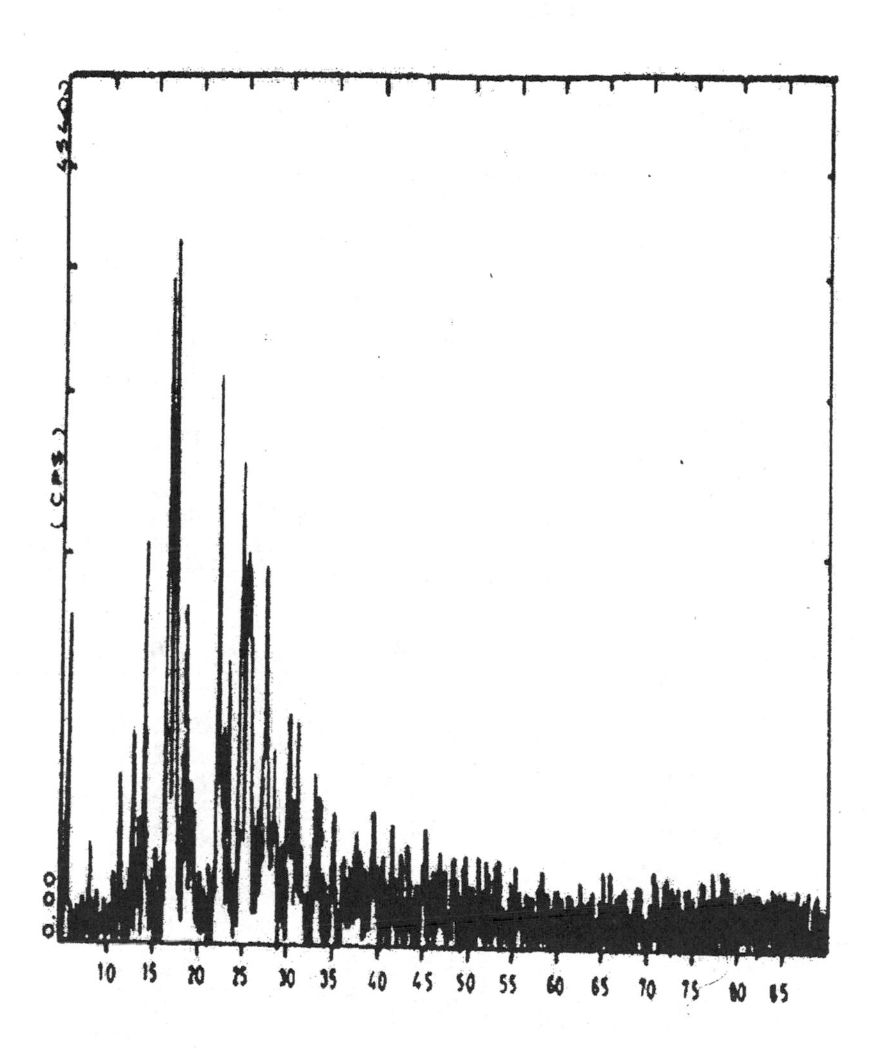

Table 1

Crystallographic Results from The X-Ray Powder Diffraction Pattern of Lansoprazole

Scattering Angle (degrees 2-)θ	d-Spacing (Å)	Relative Intensity (%)
5.651	15.6271	16.30
8.026	11.0062	1.91
11.372	7.7747	5.31
12.745	6.9400	8.80
13.272	6.6657	2.99
13.599	6.5058	3.12
14.207	6.2291	32.07
15.787	5.6090	1.23
16.610	5.3328	20.26
16.864	5.2529	88.51
17.498	5.0640	100.00
18.601	4.7663	22.53
19.197	4.6196	3.34
22.292	3.9848	64.53
22.883	3.8830	8.24
23.452	3.7902	15.73
27.954	3.5696	45.60
25.641	3.4713	29.91
25.856	3.4430	27.59
26.762	3.3285	3.20
27.410	3.2512	9.08
27.734	3.2139	27.79

Table 1 (continued)

Crystallographic Results from The X-Ray Powder Diffraction Pattern of Lansoprazole

Scattering Angle (degrees 2-)θ	d-Spacing (Å)	Relative Intensity (%)
28.551	3.1238	7.26
30.110	2.9655	7.09
30.214	2.9556	10.62
31.201	2.8642	7.64
31.638	2.8256	1.76
32.975	2.7141	4.92
33.527	2.6707	3.69
35.053	2.5578	3.13
36.071	2.4879	1.24
37.508	2.3958	1.26
39.476	2.2808	3.40
41.615	2.1684	2.08
45.150	2.0065	1.59
46.853	1.9375	1.66
49.597	1.8365	1.42
53.233	1.7193	1.36
70.867	1.3286	0.87
77.156	1.2353	1.18
81.914	1.1751	0.52

Figure 2. Differential scanning calorimetry thermogram of Lansoprazole.

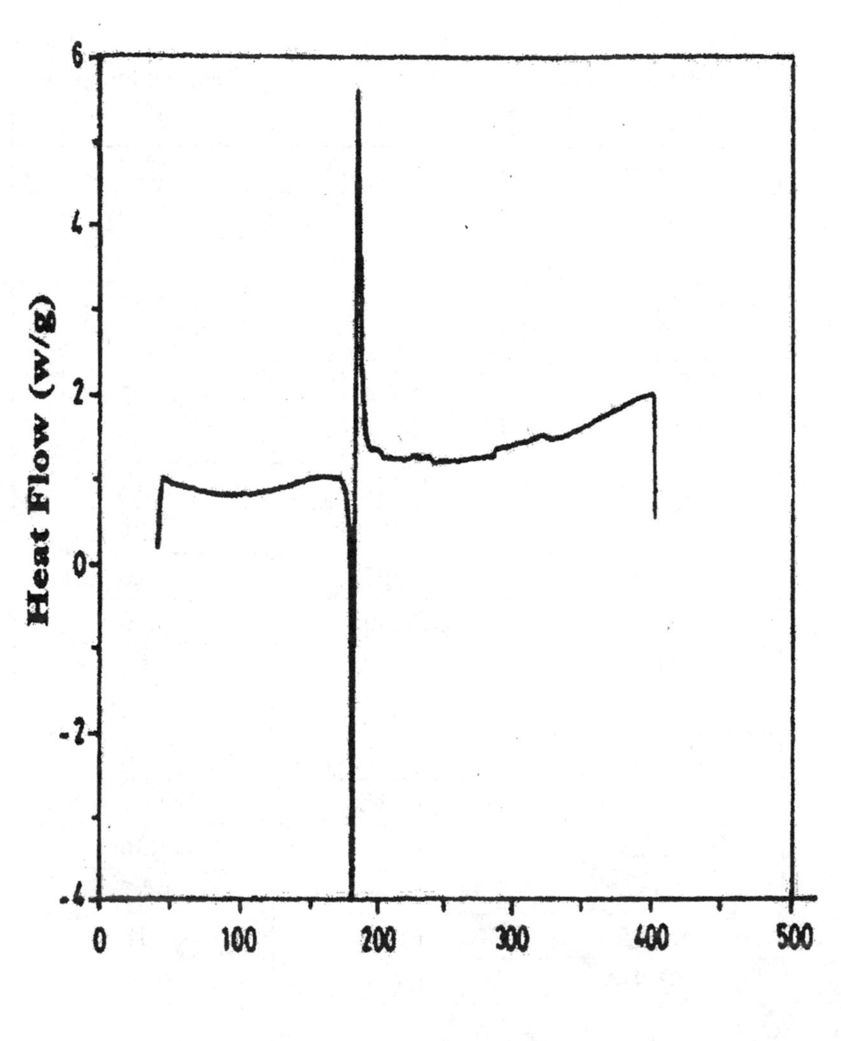

3.6 Spectroscopy

3.6.1 UV/VIS Spectroscopy

The UV spectrum of Lansoprazole was recorded using a Shimadzu model 1601-PC UV/VIS spectrophotometer. To obtain the spectrum, Lansoprazole was dissolved in methanol at a concentration of 20 μg/mL. As shown in Figure 3, the spectrum consisted of one maximum at 285 nm, characterized by $A_{1\%, 1\ cm}$ = 407.5 and molar absorptivity = 15052.

3.6.2 Vibrational Spectroscopy

The infrared absorption spectrum of Lansoprazole was obtained using a Perkin-Elmer infrared spectrophotometer, with the substance being pressed in a KBr pellet. As shown in Figure 4, the principal absorption bands were observed at 3200, 2960, 2900, and 1660 cm^{-1}.

3.6.3 Nuclear Magnetic Resonance Spectrometry

3.6.3.1 ^{1}H-NMR Spectrum

The ^{1}H NMR spectrum of Lansoprazole dissolved in $CDCl_3$ is shown in Figure 5. Confirmation of all spectral assignments was derived from a COSY experiment (which is found in Figure 6), and the assignments are summarized in Table 2.

A singlet was observed at δ = 2.22 ppm, and which integrated for the three protons associated with the aromatic methyl group. A multiplet at δ = 4.35 - 4.70 ppm (which integrated for two protons) was assigned for to the CH_2-CF_3 group, and featured obvious fluorine coupling. An AB_q pattern (J=14 Hz) was observed at δ = 4.75 - 4.87 ppm, and integrated for two protons belonging to CH_2SO. The aromatic part of the spectrum showed two doublets at δ = 6.67 and 8.35 ppm (J=5.5 Hz), assigned to the pyridine ring protons. Also observed was a multiplet at δ = 7.29-7.35 ppm belonging to the phenyl part of the structure. A broad singlet resonating at δ = 7.68 ppm was correlated to the NH group of the structure.

Figure 3. The ultraviolet absorption spectrum of Lansoprazole in methanol (20 μg/mL).

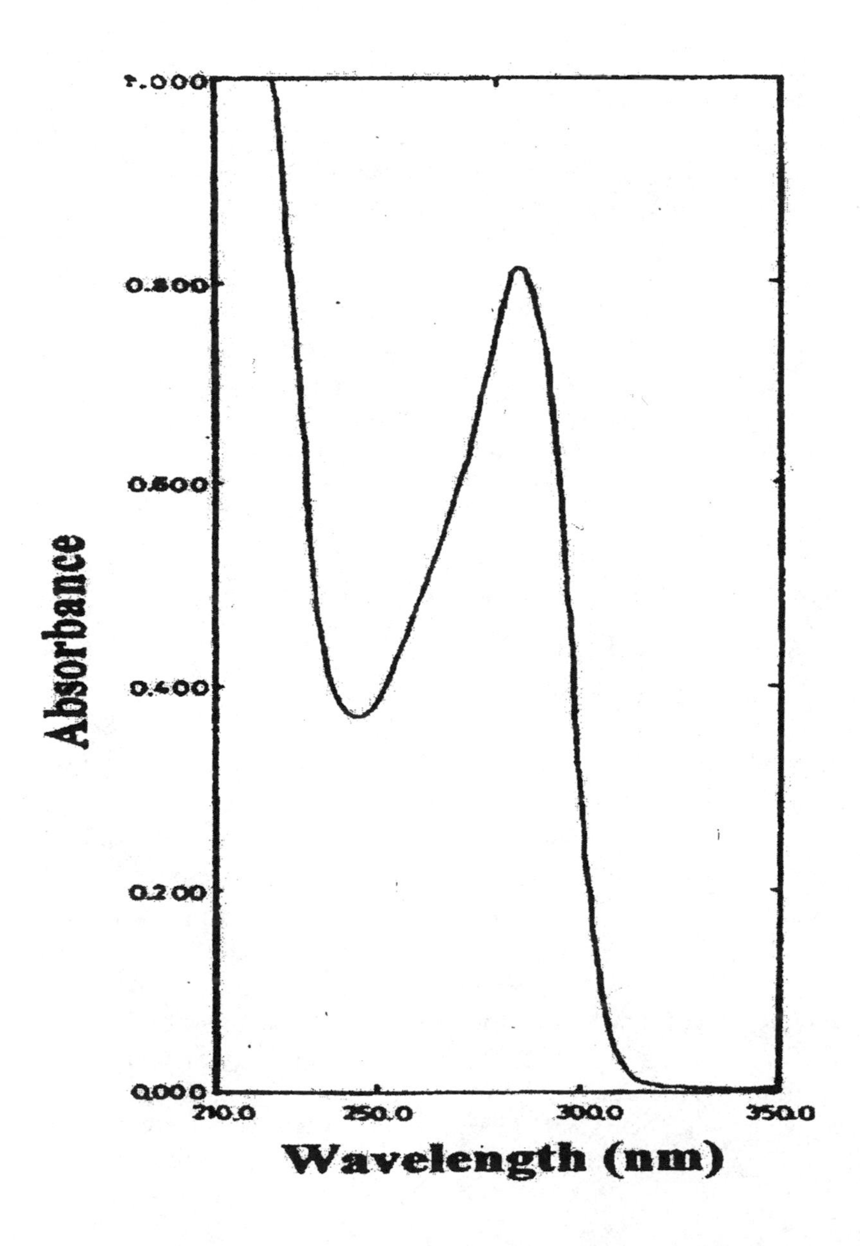

Figure 4. The infrared absorption spectrum of Lansoprazole, obtained in a KBr pellet.

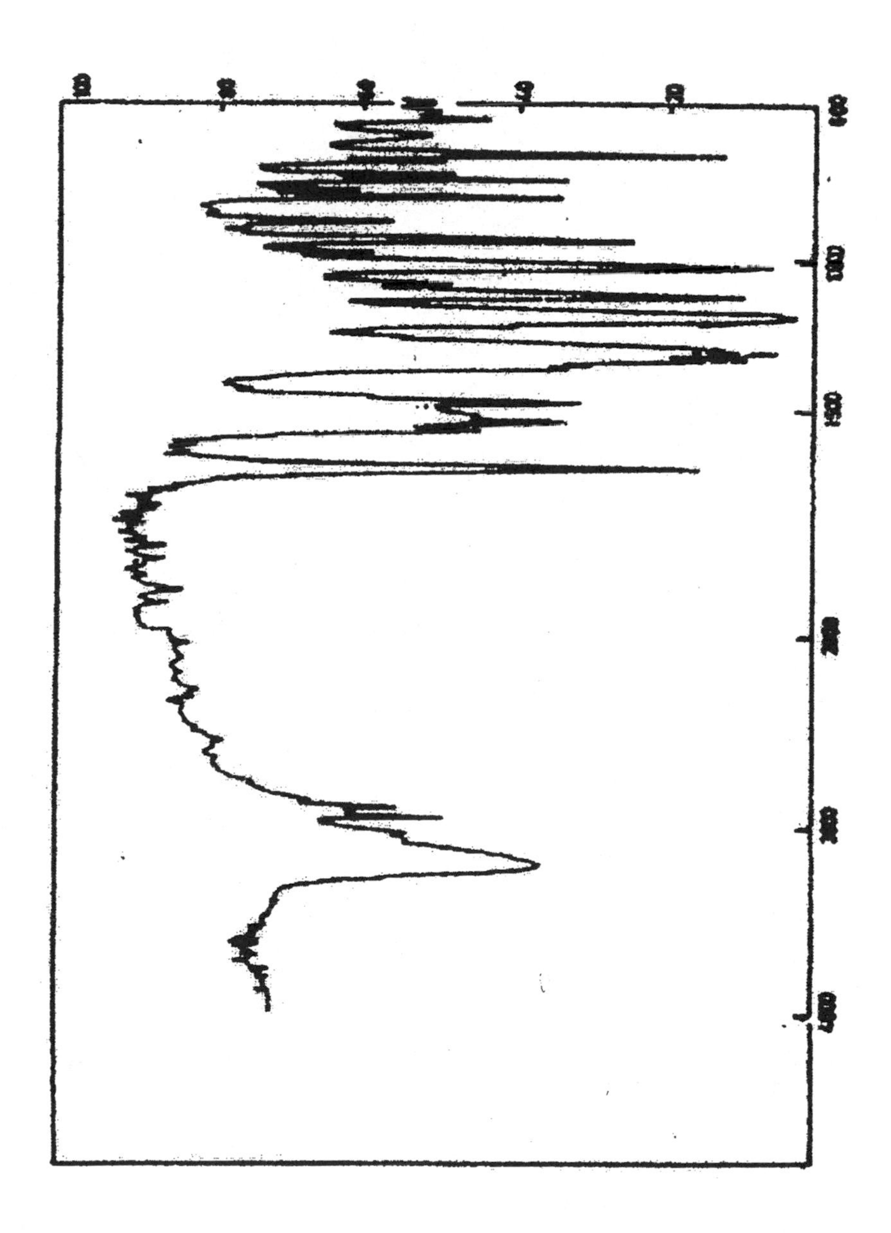

Figure 5. The ^{1}H-NMR spectrum of Lansoprazole in $CDCl_3$.

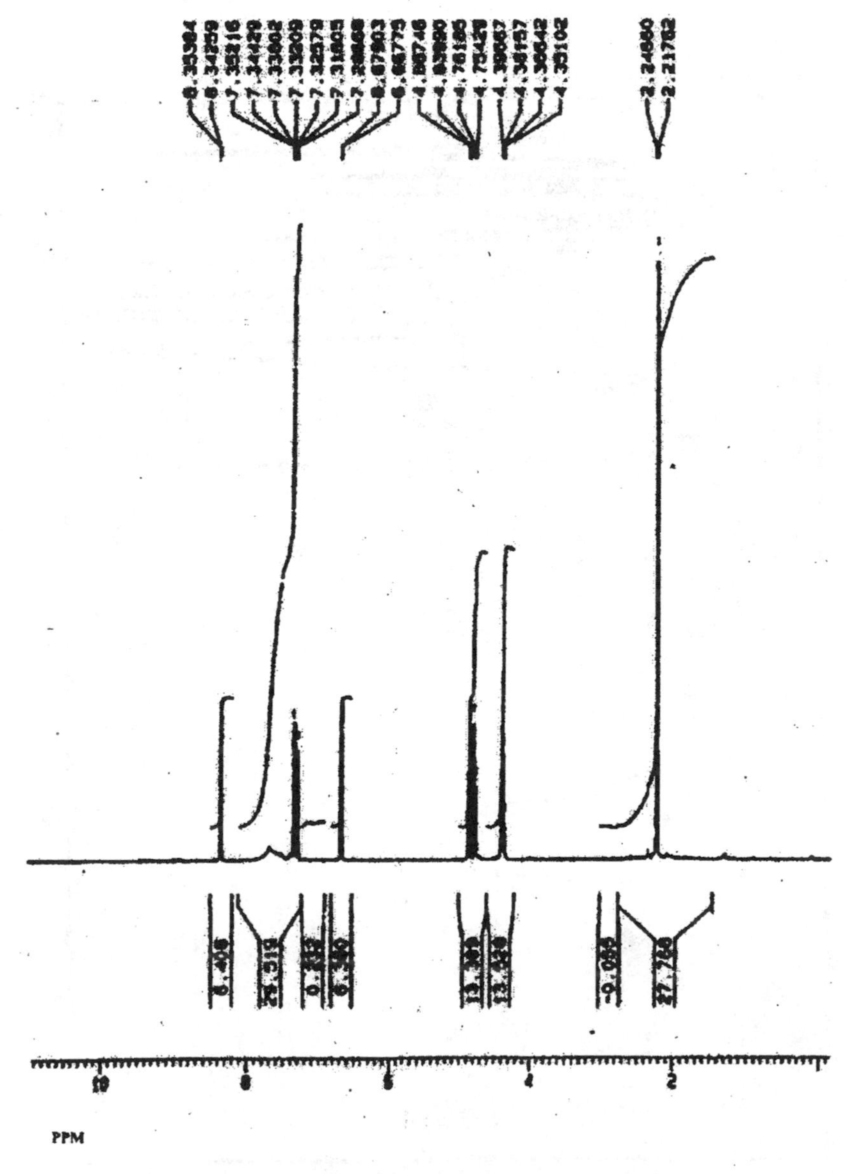

Chemical Shift (ppm)

Figure 6. The COSY ^{1}H-NMR spectrum of Lansoprazole.

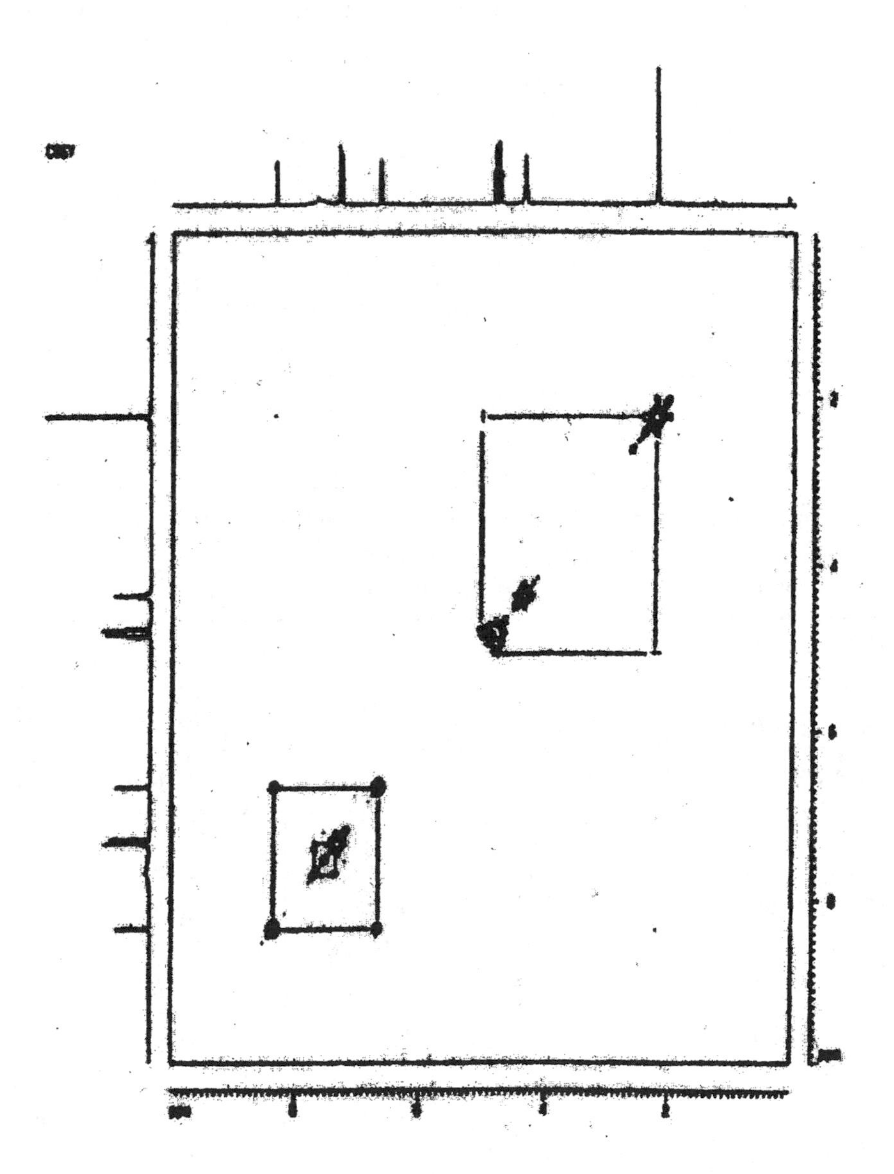

Table 2

Summary of Assignments for the ^{1}H Nuclear Magnetic Resonance Bands of Lansoprazole

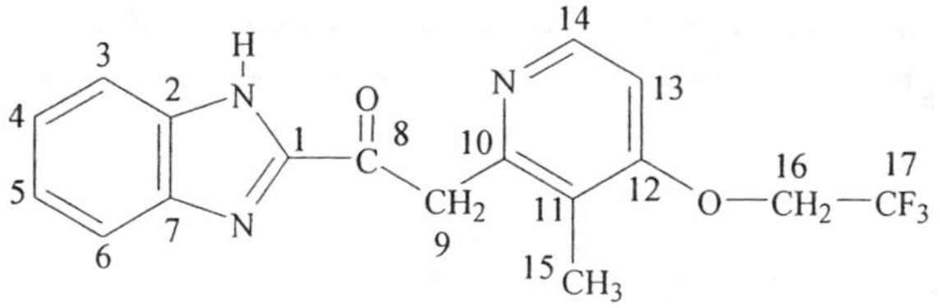

Chemical Shift, relative to TMS (ppm)	Number of protons	Multiplicity, and Coupling Constant (Hz)	Assignment
2.22	3	(5)	$\underline{CH_3}$-pyridine.
4.35-4.70	2	m	O-CH_2CF_3
4.75-4.87	2	AB_q (14)	$\underline{CH_2}$SO
6.67	1	d (5.5)	Pyridine CH
7.29-7.35	4	m	Phenyl CH
7.68	1	brs	NH
8.35	1	d(5.5)	Pyridine CH

d = doublet, m = multiplet; brs = broad singlet, AB_q = AB quartet.

3.6.3.2 ^{13}C-NMR Spectrum

The noise modulated broadband decoupled ^{13}C-NMR spectrum of Lansoprazole is shown in Figure 7, and the gradient HMQC experiment is found in Figure 8. One methyl carbon was found to resonate at $\delta = 11.3$ ppm, two CH_2 carbons at $\delta = 64.2$ and 65.7 ppm, two protonated pyridine carbons at $\delta = 106.4$ and 148.8 ppm, and two protonated phenyl carbons (owing to the symmetry at this part of structure) at $\delta = 123.8$ and 124.3 ppm. A CF_3 carbon was found to resonate at $\delta = 150.9$ ppm, in addition to another five quaternary absorptions at $\delta = 122.0$, 140.0, 142.0 153.4, and 162 ppm. The assignments are summarized in Table 3.

3.7 Mass Spectrometry

The mass spectrum of Lansoprazole was obtained utilizing a Shimadzu PQ-5000 mass spectrometer. The detailed mass fragmentation pattern is shown in Figure 9, where a base peak was observed at m/z = 238. The molecular ion peak was located at m/z = 369. The proposed mass fragmentation pattern of the drug is summarized in Table 4.

4. Methods of Analysis

4.1 Spectroscopic Analysis

Several spectrophotometric methods have been reported in the literature for the determination of Lansoprazole in its bulk and in its dosage forms.

4.1.1 Colorimetry

Mustafa reported the determination of Lansoprazole, either in pure form or in pharmaceutical formulations, by three methods [11]. Two of these methods were based on a charge transfer complexation reaction of the drug. In one case it acts as the electron pair donor complexing with a pi-acceptor (2,3-dichloro-5,6-dicyano-1,4-benzoquinone, or DDQ), and in another it complexes with the sigma acceptor iodine. The third method investigated depended on ternary complex formation with eosin and Cu(II). The colored product was quantified spectrophotometrically using

Figure 7. The noise-decoupled carbon-^{13}C-NMR spectrum of Lansoprazole.

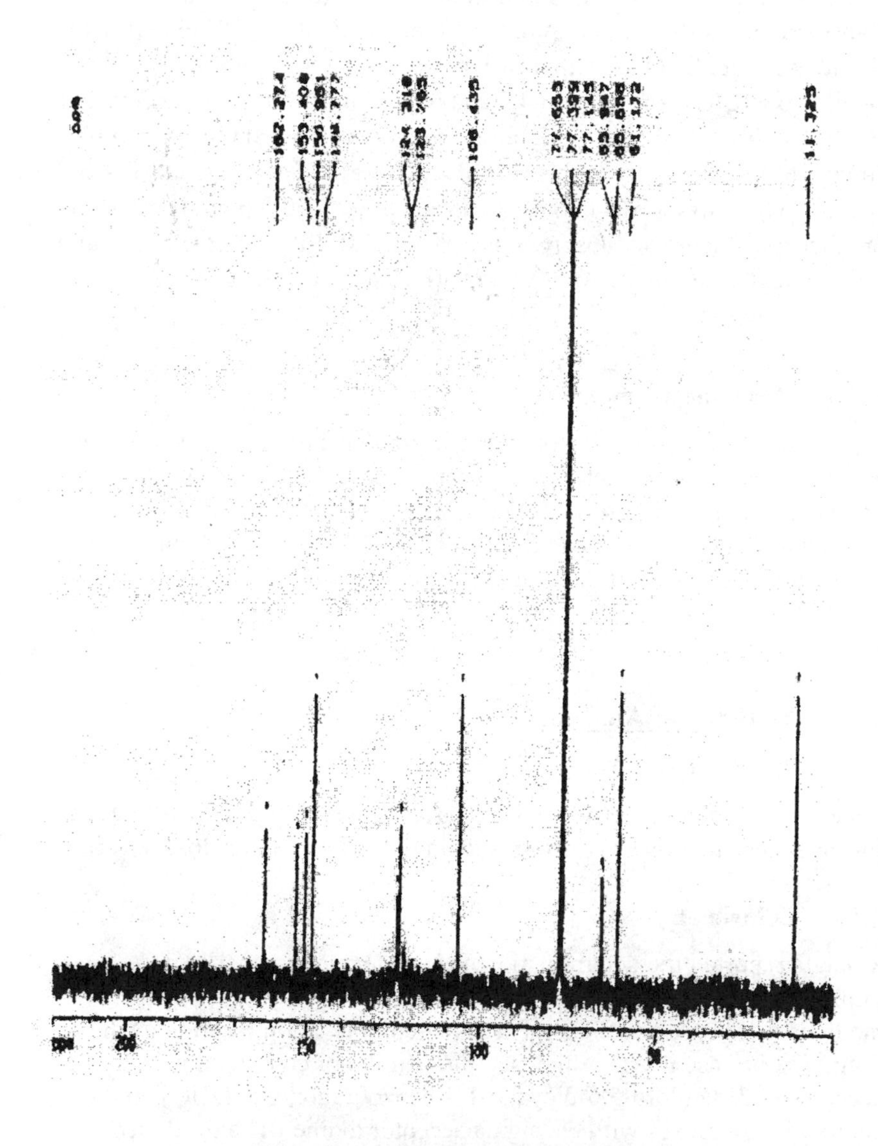

Chemical Shift (ppm)

Figure 8. The HMQC ^{13}C-NMR spectrum of Lansoprazole.

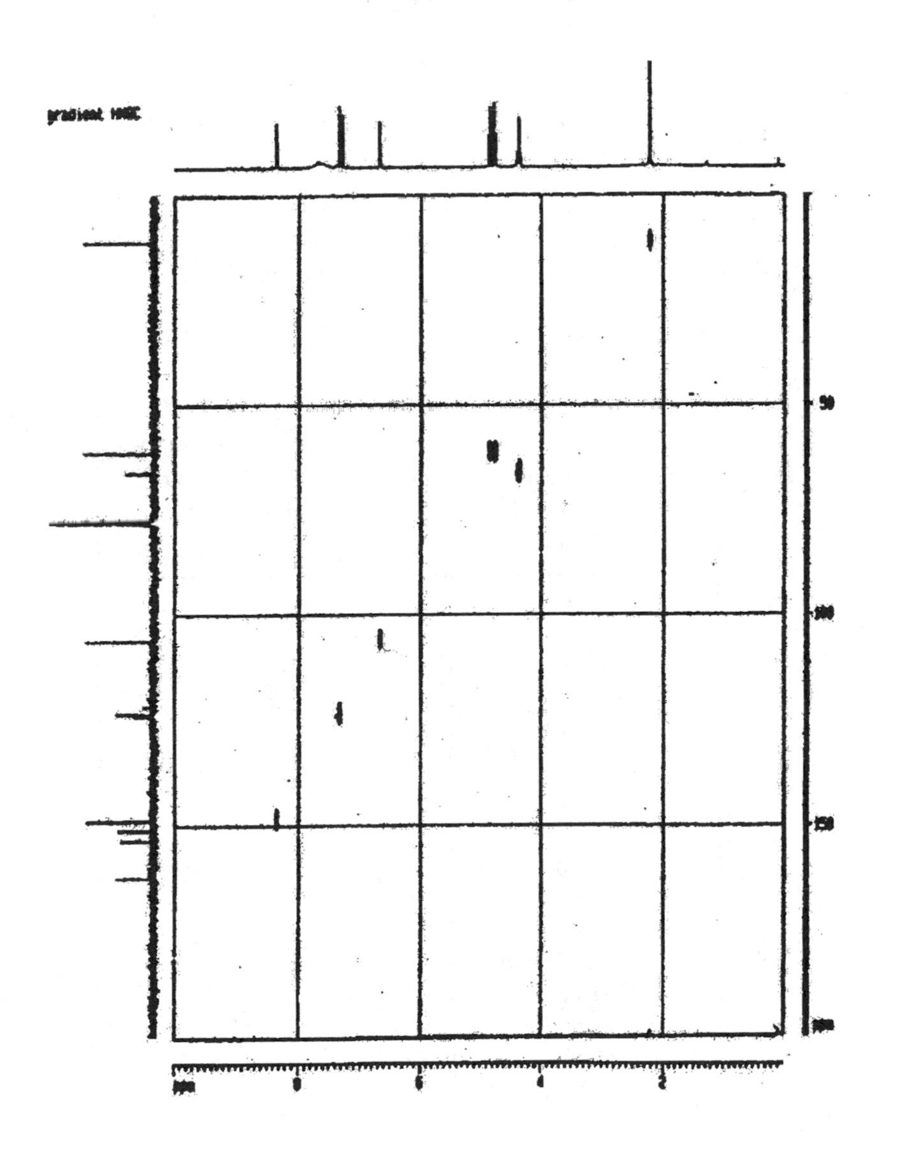

Table 3

Summary of Assignments for the ^{13}C Nuclear Magnetic Resonance Bands of Lansoprazole

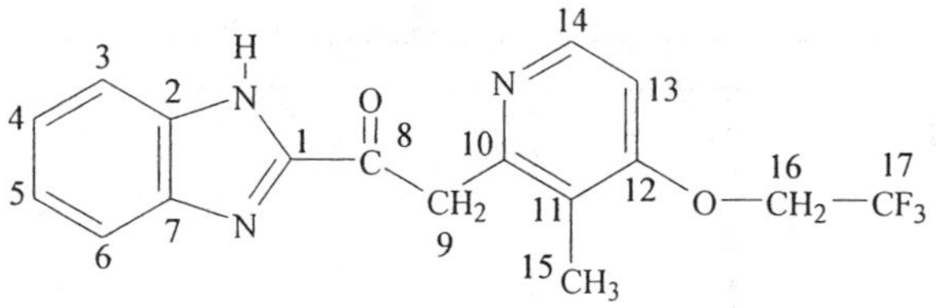

Chemical Shift, relative to TMS (ppm)	Assignment (Carbon Number)
11.3	15
64.2	16
65.7	9
106.4	3 + 6
123.8	4 + 5
124.3	13
148.8	17
150.9	1
153.4	8

d = doublet, m = multiplet; brs = broad singlet, AB_q = AB quartet.

Figure 9. The mass spectrum of Lansoprazole.

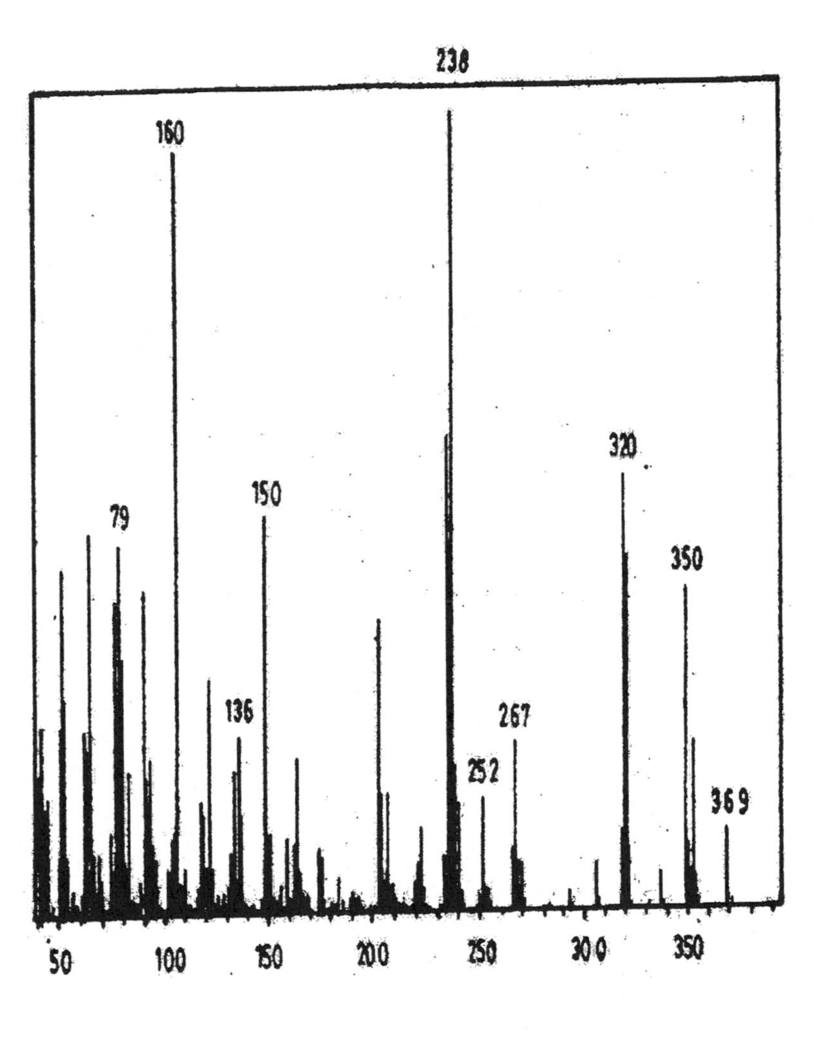

Table 4

Summary of Assignments for the Fragmentation Ions Observed in the Mass Spectrum of Lansoprazole

m/z	Relative intensity	Fragment
369	9%	H N O S–CH_2 N OCH_2CF_3 CH_3]+•
350	39%	H N O S–CH_2 N OCH_2CF_2 CH_3]+•
320	52%	H N O S–CH= H_2C O–CH=CF_2]+•
319	9%	N^+ O S–CH= H_2C O–CH=CF_2
293	2%	H N O S–CH= H_3C O–CH_2–CF_3]+•
268	6%	H N O S–CH= H_2C O–CH=CF_2]+•

Table 4 (continued)

Summary of Assignments for the Fragmentation Ions Observed in the Mass Spectrum of Lansoprazole

267	20%	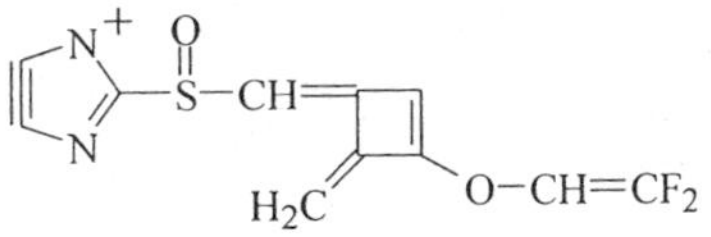 N^+ O S–CH= N H_2C O–CH=CF_2
252	14%	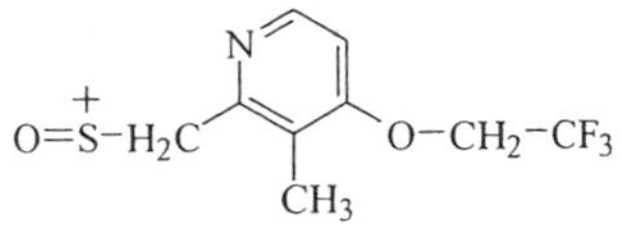 N + O=S–H_2C O–CH_2–CF_3 CH_3
238	100%	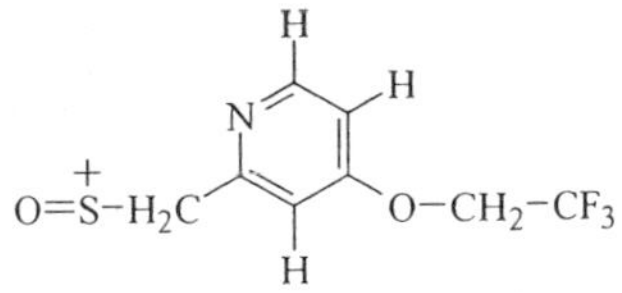 H N H + O=S–H_2C O–CH_2–CF_3 H
204	34%	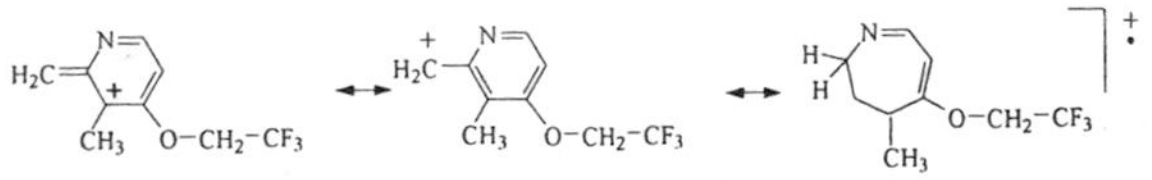 N H_2C= + CH_3 O–CH_2–CF_3 ↔ + N H_2C– CH_3 O–CH_2–CF_3 ↔ H N H O–CH_2–CF_3 CH_3]+•
185	4%	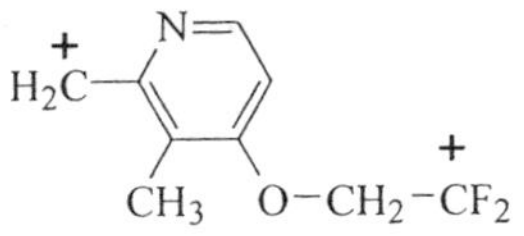 + N H_2C– + CH_3 O–CH_2–CF_2
175	8%	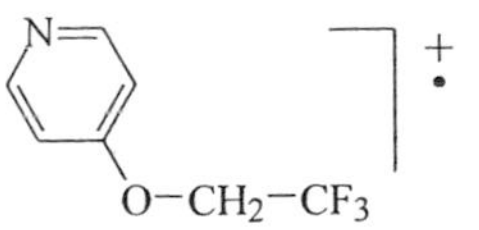 N O–CH_2–CF_3]+•
165	19%	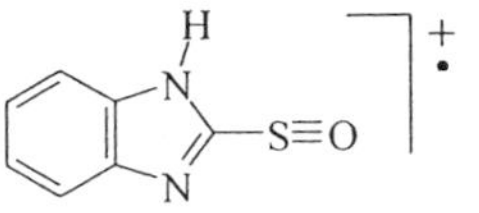 H N –S≡O N]+•

Table 4 (continued)

Summary of Assignments for the Fragmentation Ions Observed in the Mass Spectrum of Lansoprazole

150	48%	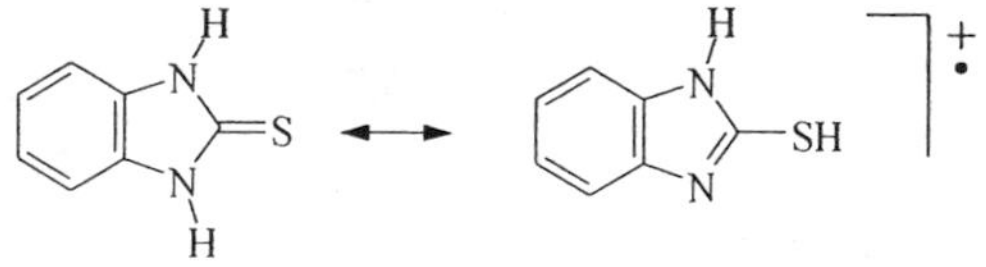
122	28%	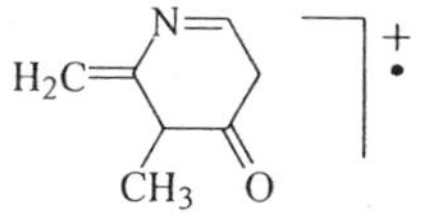
106	93%	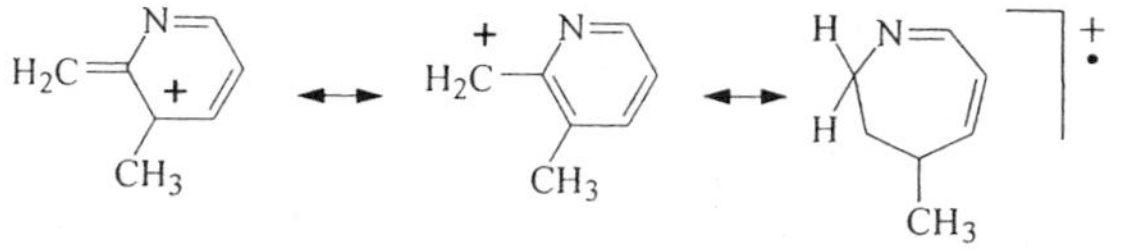
90	40%	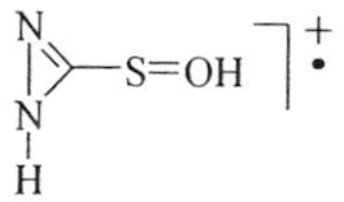
79	45%	$HO^{+}=C=CF_2$
65	45%	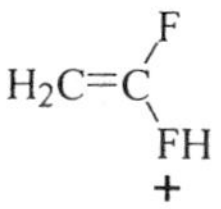 $H_2C=C(F)FH^{+}$
52	42%	

absorption bands at 457 nm for DDQ (method 1), at 293 and 359 nm for iodine (method 2), and at 549 nm using ternary complex formation (method 3). These methods have enabled the determination of Lansoprazole in concentration ranges from 10-90, 1.48-6.65, and 3.69-16.61 μg/mL.

Meyyanathan *et al.* described two spectrophotometric methods for the determination of Lansoprazole in its dosage forms [12]. In the first method, the powdered contents of Lansoprazole capsules (200 mg) were dissolved in 50 mL of dimethylformamide (DMF), filtered through sintered glass, and diluted to 100 mL with DMF. One milliliter of this solution was mixed with 0.4 mL of acetyl chloride, 1 mL of 1% of cupper sulfate, and the volume made up to 10 mL with DMF. The resulting yellow color was measured at 478.5 nm, and was stable for three hours. Beer's law was obeyed over the range of 100-600 μg/mL.

In the second method of Meyyanathan *et al.*, the powdered capsule contents were dissolved in 50 mL methanol, filtered through sintered glass, and diluted to 100 mL with methano [12]l. One milliliter of this solution was mixed with 0.4 mL of 0.3% 3-methyl-2-benzthiazolinone hydrazone and 1 mL of ceric ammonium sulfate. The mixture was heated at 100°C for 5 minutes on a water-bath, cooled, and diluted to 10 mL with water. The resulting red color was measured at 491.2 nm, and was stable for 90 minutes. Beer's law was obeyed over the range of 100-500 μg/mL.

4.1.2 Spectrophotometry

Ozaltin developed two different methods based on ultraviolet absorption spectroscopy for the determination of Lansoprazole in pharmaceutical dosage forms [13]. The solution of the standard substance and the sample were prepared in 0.1 M NaOH and phosphate buffer pH 6.6. Both total absorption and derivative spectroscopic techniques were applied. The linear range for the UV-spectrophotometric method was 3-25 μg/mL, and 0.5-25 μg/mL for the derivative spectroscopic method. The methods were applied to three different pharmaceutical preparations, with recoveries being equal to 100.2%.

4.2 Chromatographic Methods of Analysis

4.2.1 Thin Layer Chromatography

A rapid and sensitive high performance thin layer chromatographic (HP-TLC) assay method for the measurement of Lansoprazole in human plasma has been developed by Pandya *et al* [14]. The method involves a single stage extraction procedure without the use of an internal standard. A known amount of drug extract was spotted on pre-coated Silica gel 60 F_{254} plates using a mobile phase consisting of 15:1 chloroform-methanol, and the Lansoprazole spots were observed by fluorescence upon excitation at 286 nm. The total area of each spot was determined with the help of a Camag TLC-Scanner 3.

Argekar and Kunjir reported a stability indicating high performance thin-layer chromatographic method for the quantification of Lansoprazole in its bulk form and in pharmaceutical preparations [15]. The contents of Lansoprazole capsules were sonicated with alkaline methanol for 15 minutes, and then diluted with methanol. After filtering, the filtrate was applied to silica gel 60 F_{254} HP-TLC plates, with the organic layer from a mixture of ethyl acetate-methanol-13.5 M ammonia solution (10:3:4) being used as the mobile phase. The plates were dried, and then scanned by densitometry at 288 nm in the absorbance-reflectance mode. Calibration graphs were linear for 40-300 ng of Lansoprazole; with a detection limit of 10 ng and recoveries of 99.56 to 100.1%.

4.2.2 High Performance Liquid Chromatography

A number of high-performance liquid chromatography (HPLC) methods have been used for the determination of Lansoprazole [16-26]. A summary of the HPLC characteristics of these methods is found in Table 5.

4.2.3 Capillary Electrophoresis

The enantiomers of Lansoprazole and other structurally related drugs were separated by a capillary zone electrophoresis, using bovine serum albumin (BSA) as the chiral selector [27]. The separations were carried out using a fused-silica column (60 cm × 50 μm i.d., 50 cm to detector), with a buffer consisting of 40 μM-BSA and 7% 1-propanol in 10 mM potassium

Table 4

Summary of HPLC Methods Used for the Determination of Lansoprazole

Column	Mobile phase	Detection and λ_{max}	Sample	Ref.
Chiral-AGP with covalently bound acid α glycoprotein.	10 mL Propan-2-ol/ 100 mL 0.1 M sodium dihydrogen phosphate/100 μl glacial acetic acid (pH 7) and made up with water to 1 liter and filtered.	UV at 283 nm.	Separation of the enantiomers in human serum.	16
Nucleosil 100 C18 (25 cm × 4 mm i.d.), (5 μm)	Acetonitrile / H_2O / n-octylamine (400:600:1) pH = 7 (0.9 mL/min)	UV at 283 nm.	Analyses in human serum.	17
C18 Spherisorb	Acetonitrile/0.1 M sodium acetate (40:60) pH = 7.	UV at 230 nm	In biological fluids and in dosage form.	18
C18 (25 cm × 4.6 mm i.d.) at 40°C (5 μm).	38% aqueous acetonitrile with 1% n-octylamine (1 mL/min)	UV at 285 nm.	In human plasma	19
(25 cm × 4.6 mm) of Nucleosil C18 (5 μm)	Acetonitrile/H_3PO_4/ 0.25 M KH_2PO_4/ acetic acid/ triethylamine/ H_2O, (2 mL/min)	UV at 285 nm and 308 nm	In plasma using a loop column.	20

Table 4 (continued)

Summary of HPLC Methods Used for the Determination of Lansoprazole

TSK gel ODS-120T (25 cm × 4.6 mm), (5 μm)	Water: acetonitrile: octylamine (620:380:1) for serum (1 mL/min)	In serum at 285 nm, and in urine at 303 nm	In human serum and urine	21
Octyldecyl-silane (15 cm × 4.6 mm id), (5 μm)	35% aqueous acetonitrile; 1mL n-octylamine; 5 mM N-acetohydroxamic acid, adjusted to pH 7 with 85% H_3PO_4 (1 mL/min)	Ultraviolet at 285 nm	The drug and five of its metabolites in plasma.	22
Chiralpak AD (an amylose-based chiral stationary phase)	Ethanol-hexane (1:4)(1 mL/min).	–	Separation of enantiomers of the drug by HPLC	23
A 10 μm chiralcel OD-R (25 cm × 4.6 mm i.d.) operated at 40°C.	Methanol-50 mM-$NaClO_4$ (13:7) (0.5 mL/min).	Ultraviolet at 286 nm	Separation of enantiomers of the drug by HPLC using cellulose-based chiral stationary phases in the reversed-phase mode.	24

Table 4 (continued)

Summary of HPLC Methods Used for the Determination of Lansoprazole

1. Chiralcel OD. 2. Chiralcel OF 3. Chiralcel OG 4. Chiralcel OJ 5. Chiral pak AS	–	–	Assay of the enantiomers of the drug using chiral stationary phase LC in human serum	25
Novapak C_{18} (4 μm) (150 × 3.9 mm id)	58% Sodium dihydrogen phosphate (0.02 M), 23% acetonitrile and 19% methanol, pH 7.3 (1.8 mL/min).	At 285 nm.	Human serum	26

phosphate of pH 7.4. Electrokinetic injection was at 5-8 kV for seven seconds. An applied voltage of 100 V/cm was used, and the detection was made at 290 nm. Detection limits were at 0.04 mg/mL.

The analysis of Lansoprazole and other pyridinyl methyl sulfinyl benzimidazoles using non-aqueous capillary electrophoresis was studied by Tivesten *et al* [28]. Using polar organic solvents as background electrolytes, optimum separation of these drugs was obtained on fused silica capillaries (48.5 × 50 μm i.d.; 40 cm effective length). The system was operated at an applied voltage of +20 kV, with 80 mM electrolyte and detection at 305 nm.

5. Stability

Lansoprazole is known to be unstable in acidic media [3].

6. Pharmacology, Metabolism, and Pharmacokinetics

Lansoprazole is a new drug used for treatment of the disorders of gastric acid hypersecretion. Lansoprazole and the other drugs in this class, such omeprazole, pantoprazole, and rabeprazole, share a common mechanism of action. Each is a substituted benzimidazole, which inhibits activity of the H^+, K^+-ATPase located on the optical surface of parietal cells by preventing the secretion of gastric acid.

Lansoprazole is rapidly absorbed after oral administration, with peak plasma concentrations being achieved about 1.5 hours after an oral dose. The bioavailability of the drug is reported to be 80% or more, even with the first dose. Since Lansoprazole is unstable at acidic pH values, it must be given in an enteric coated dosage form. It is extensively metabolized in the liver and excreted primarily in feces via the bile, with only about to 30% a dose being excreted in urine [2,3].

The plasma elimination half-life of the drug is around 1.4 to 2 hours, but the duration of action is much longer. Lansoprazole is about 97% bound to plasma protein. Clearance is decreased in elderly patients, and in those with liver disease [29,30]. Lansoprazole dose is given by mouth as an enteric-coated formulation, and is normally taken before food in the morning. The dosing is 15-30 mg twice daily, by mouth, for 2-4 weeks [3, 30-32].

A liquid formulation of Lansoprazole suppressed intra-gastric acidity when given through a gastrostomy [33]. *Helicobacter pylori* eradication significantly decreased the intra-gastric pH reached during Lansoprazole treatment throughout all periods [34].

Acknowledgment

The authors would like to thank Mr. Tanvir A. Butt, Department of Pharmaceutical Chemistry, College of Pharmacy, King Saud University for his technical assistance.

References

1. ***The Merck Index***, 12th edn., S. Budavari, ed., Merck and Co., Inc., Rahway, NJ, 1996, p. 916.
2. E.A. Lew, *Aliment. Pharmacol. Therapap.*, **13**, 11 (1999).
3. ***Martindale, The Extra Pharmacopeia***, 32nd edn., J.E.F. Reynold, ed., Royal Pharmaceutical Society, London, 1999, p. 1196.
4. L.B. Barradell, D. Faulds, and D. McTavish, *Drugs*, **44**, 225 (1992).
5. H. Satoh, N. Inatomi, H. Nagaya, I. Inada, A. Nohara, N. Nakamura, and Y. Maki, *J. Pharmacol. Exp. Therap.* **248**, 806 (1989).
6. D.C. Metz, J.R. Pisegna, G.L. Ringham, K. Feigenbaum, P.D. Kovialk, P.N. Maton, J.D. Gardner, and R.T. Jensen, *Dig. Dis. Sci.*, **38**, 245 (1993).
7. W.R. Garnett, *Ann. Pharmacotherap.*, **30**, 1425 (1996).
8. M. Inoue and M. Nakamura, *J. Clin. Gastroenterol.*, **20**, 17 (1995).

9. K. Kubo, K. Oda, T. Kaneko, H. Satoh, and A. Nohara, *Chem. Pharm. Bull.*, **38**, 2853 (1990).

10. D. Lednicer, in ***The Organic Chemistry of Drug Synthesis***, Vol. 5, John Wiley & Sons, Inc., New York, 1995, 115.

11. A.A. Moustafa. *J. Pharm. Biomed. Anal.*, **22**, 45 (2000).

12. S.N. Meyyanathan, J.R.A. Raj, and B. Suresh. *Indian Drugs*, **34**, 403 (1997).

13. N. Ozaltin, *J. Pharm. Biomed. Anal.*, **20**, 599 (1999).

14. K.K. Pandya, V.D. Mody, M.C. Satia, A.I. Modi, R.I. Modi, B.K. Chakravarthy, and T.P. Gandhi. *J. Chrom. Biomed. Sci.*, **693**, 199 (1997).

15. A.P. Argekar and S.S. Kunjir, *J. Planar Chrom. Mod. TLC*, **9**, 296 (1996).

16. K. Borner, E. Borner, and H. Lode, *Chromatographia*, **47**, 171 (1998).

17. K. Borner, E. Borner, and H. Lode, *Chromatographia*, **45**, 450 (1997).

18. Avgerinos, Th. Karidas, C. Potsides, and S. Axarlis, *Eur. J. Drug Metab. Pharmacokinet.*, **23**, 329 (1998).

19. Y.M. Li, L.Y. Chen, L.J. Ma, and Q.Y. Zhang, *Yaowu-Fenxi-Zazhi*, **16**, 252 (1996).

20. B.D. Landes, G. Miscoria and B. Flouvat, *J. Chrom. Biomed. Appl.*, **577**, 117 (1992).

21. Aoki; M. Okumura and T. Yashiki, *J. Chrom. Biomed. Appl.*, **571**, 283 (1991).

22. D.M. Karol; G.R. Granneman, and K. Alexander, *J. Chrom. Biomed. Appl.*, **668**, 182 (1995).

23. K. Balmer, B.A. Persson, and P.O. Lagerstroem, *J. Chrom.*, **660**, 269 (1994).

24. M. Tanaka, H. Yamazaki, and H. Hakushi, *Chirality*, **7**, 612 (1995).

25. H. Katsuki, H. Yagi, K. Arimori, C. Nakamura, M. Nakano, S. Katafuchi, Y. Fujioka, and S. Fujiyama, *Pharm. Res.*, **13**, 611 (1996).

26. M.F. Zaater, N. Najib, and E. Ghanem, *Saudi Pharmaceutical Journal*, **7**, 123 (1999).

27. D. Eberle, R.P. Hummel and R. Kuhn, *J. Chrom.*, **759**, 185 (1997).

28. Tivesten, S. Folestad, V. Schonbacher, and K. Svensson, *Chromatographia*, **49**, 57 (1999).

29. H. Harder, S. Teyssen, F. Stephan, R. Pfutzer, G. Kiel, W. Fuchs, and M.V. Singer, *Scand. J. Gastroenterol.*, **34**, 551 (1999).

30. V.K. Sharma, A.E. Ugheoke, R. Vasudeva, and C.W. Howden, *Aliment Pharmacol. Therap.*, **12**, 1171 (1998).

31. M. Sugiyama, T. Aoki and Y. Matsuo, *J. Clin. Gastroenterol.*, **20** 14 (1995).

32. K. Kihira, Y. Yoshida, T. Kasano, Y. Taniguchi, K. Sato, K. Kimura, M. Hirose, and H. Koyama, *Nippon Shokakibyo Gakkai Zasshi*, **88**, 672 (1991).

33. V.K. Sharma, R. Vasudeva, and C.W. Howden, *Am. J. Gastroenterol.*, **94**, 1813 (1999).

34. M.A. Van Herwaarden, M. Samson, CH van Nipsen, P.G. Mulder, and A.J. Smout, *Aliment Pharmacol. Therap.*, **13**, 731 (1999).

MALIC ACID

Harry G. Brittain

Center for Pharmaceutical Physics
10 Charles Road
Milford, NJ 08848
USA

1075-6280/01 $35.00

Contents

1. Description

1.1 Nomenclature

1.1.1 Systematic Chemical Name

1-Hydroxy-1,2-ethanedicarboxylic acid

Hydroxybutanedioic acid

Hydroxysuccinic acid

1.1.2 Nonproprietary Names

Malic acid; Apple acid (L-enantiomer)

1.2 Formulae

1.2.1 Empirical Formula, Molecular Weight, CAS Number

$C_4H_6O_5$ [MW = 134.087]

1.2.2 CAS Numbers

Racemate: 617-48-1

L-enantiomer: 97-67-6

D-enantiomer: 636-61-3

1.2.3 Structural Formula

O OH

HO O

OH

1.3 Elemental Analysis

The calculated elemental composition is as follows:

carbon:	35.83%
hydrogen:	4.51%
oxygen:	59.66%

1.4 Appearance

Malic acid is obtained as a white, or nearly white, crystalline powder or granules having a slight odor and a strongly acidic taste [1]. The synthetically produced material is a racemic mixture, while the naturally occurring substance is levorotatory (the L-enantiomer).

1.5 Uses and Applications [2]

Malic acid is used an acidulant, an antioxidant flavoring agent, a buffering agent, and a chelating agent. In pharmaceutical formulations, the substance is used as a general-purpose acidulant. The (*L*)-enantiomer possesses a slight apple flavor, and is used as a flavoring agent to mask bitter tastes and to provide tartness. Malic acid is also used as an alternative to citric acid in effervescent powders, mouthwashes, and tooth-cleaning tablets. In addition, it has chelating and antioxidant properties and may be used as a synergist, with butylated hydroxytoluene, to retard oxidation in vegetable oils.

Therapeutically, malic acid has been used topically in combination with benzoic acid and salicylic acid to treat burns, ulcers, and wounds. It has also been used orally and parenterally, either intravenously or intramuscularly, in the treatment f liver disorders, and as a sialagogue.

2. Method(s) of Preparation

The (L)-enantiomer is naturally occurring, and has been found in apples and other fruits and plants such as grapes, gooseberries, and rhubarb stalks [3]. The substance was first isolated by Scheele in 1785 from unripe apples [4]. Calcium malate separates during the concentration of maple sap, and is known as *sugar sand*.

Racemic malic acid is manufactured by the hydration of maleic acid [4]:

$$\xrightarrow{H_2O}$$

It can also be produced by the hydration of fumaric acid [4]:

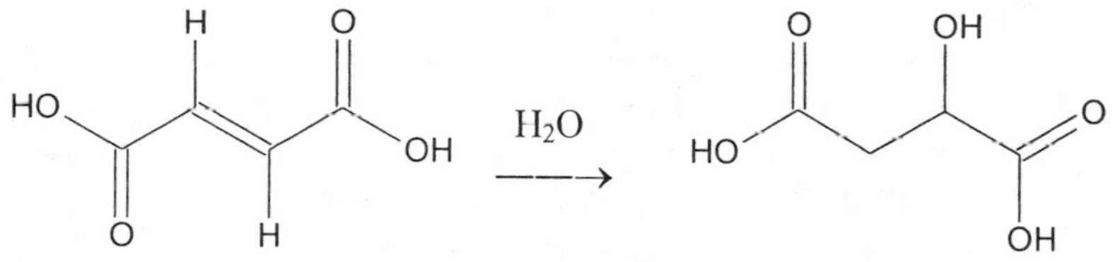

$$\xrightarrow{H_2O}$$

Preparation of the enantiomerically pure and racemic forms has been described in the literature [5]. Either (*L*)-malic acid or (*D*)-malic acid can be obtained enantiomerically pure through resolution of the racemic mixture, and methods for such purification are available [6]. The typical resolution uses either quinine or cinchonidine as the resolving agent.

3. Physical Properties

3.1 Ionization Constants

Citric acid is a moderately strong organic acid. When measured at 25°C and an ionic strength of 0.1, the pKa values are $pK_1 = 4.71$ and $pK_2 = 3.24$ [7]. As would be expected, the ionization constants decrease when the ionic strength is raised to 1.0 ($pK_1 = 4.45$ and $pK_2 = 3.11$), and increase when the ionic strength is decreased to 0 ($pK_1 = 5.097$ and $pK_2 = 3.459$).

At zero ionic strength, the enthalpy change (ΔH) associated with the two ionizations has been reported to be 0.28 and –0.71 kcal/mol, and the entropy change is 24.3 and 13.5 EU [7].

3.2 Metal Ion Binding Characteristics

Malic acid is a well-known chelating agent for a wide variety of metal ions [7]. A summary of reporting binding constants is given in Table 1.

3.3 Solubility Characteristics

The solubility of both the optically active and racemic forms of malic acid have been reported in a number of solvents [3]. A summary of the reported data is given in Table 2.

Table 1

Binding Constants, Measured at 25°C, of Malic acid with Various Monovalent and Divalent Metal Ions

Metal Ion	log K_1	log K_2
Li(I)	0.38	–
Na(I)	0.28	–
K(I)	0.18	–
Mg(II)	1.70	0.90
Ca(II)	1.96	1.06
Ba(II)	1.45	0.67
Mn(II)	2.24	not reported
Fe(II)	2.6	not reported
Co(II)	2.86	1.64
Ni(II)	3.17	1.83
Cu(II)	3.42	2.00
Zn(II)	2.93	1.66

Note: All results were obtained at an ionic strength of 0.1

Table 2

Solubilities (at 20°C) of Malic acid in Various Solvents

	Concentration (g solute per 100 g solvent	
Solvent	(*DL*)-Malic Acid	(*L*)-Malic Acid
Water	55.8	36.35
Methanol	82.70	197.22
Ethanol	45.53	86.60
Dioxane	22.70	74.35
Acetone	17.75	60.66
Diethyl Ether	0.84	2.70
Benzene	practically insoluble	practically insoluble

The temperature dependence of the aqueous solubility of malic acid has been reported [7], and these data are plotted in Figure 1. For a solution having a concentration of 10.4 mol/kg, the apparent molar enthalpy of solution at 298.15 K was determined to be 12.7 kJ/mol. This value differed from the value of 21.8 kJ/mol obtained using calorimetry at a lower concentration (0.0318 mol/kg and 298.15 K), indicating that the enthalpy of dilution was important in the calorimetric determination.

Using the ACD PhysChem program (Advanced Chemistry Development, Toronto CA), the aqueous solubility of (*DL*)-malic acid was calculated to be approximately 62.5 g solute per 100 g solvent [8]. This value agrees fairly well with the literature value quoted in Table 2.

3.4 Partition Coefficients

Using the ACD PhysChem program (Advanced Chemistry Development, Toronto CA), the octanol-water partition coefficient of (*DL*)-malic acid was calculated to be –1.26 [8], demonstrating the hydrophilic nature of the compound. As evident in Figure 2, malic acid becomes even more hydrophilic once it becomes ionized.

3.5 Optical Activity

Malic acid contains a single center of dissymmetry, and hence is capable of existing as the naturally occurring (*L*)-enantiomer, the mirror image (*D*)-enantiomer, and the racemic mixture of these.

The absolute configuration of (*L*)-malic acid has been determined by the absolute x-ray diffraction method, and has been found to be of the (*S*)-configuration [9].

3.5.1 Optical Rotation

The specific rotation of (*L*)-malic acid has been reported at 589 nm as –2.3° (c = 8.5) [3], and the substance is levorotatory in most solvents. However, the specific rotation is known to exhibit a strong dependence upon solvent, leading to a sign reversal in some solvent systems [10]. Formation of metal salts also strongly perturbs the optical rotation of malic acid, with the uranyl ion exerting one of the most profound effects [11].

Figure 1. Temperature dependence of the aqueous solubility of (*DL*)-malic acid.

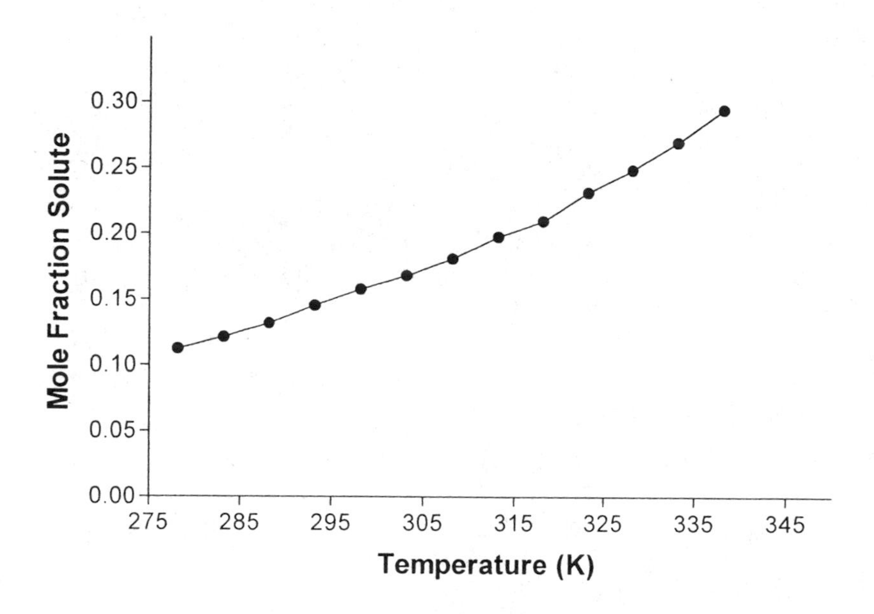

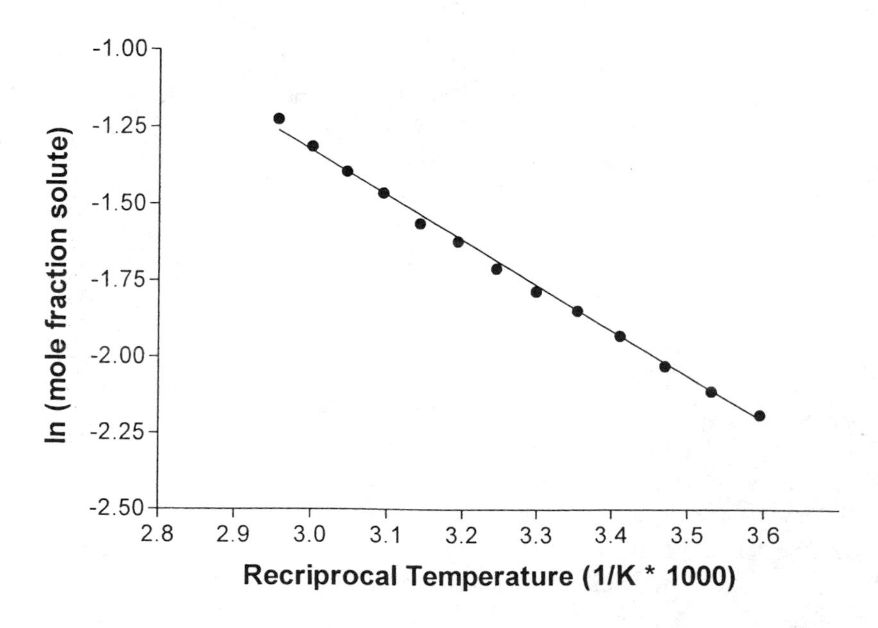

Figure 2. pH dependence of the calculated distribution coefficient of (*DL*)-malic acid.

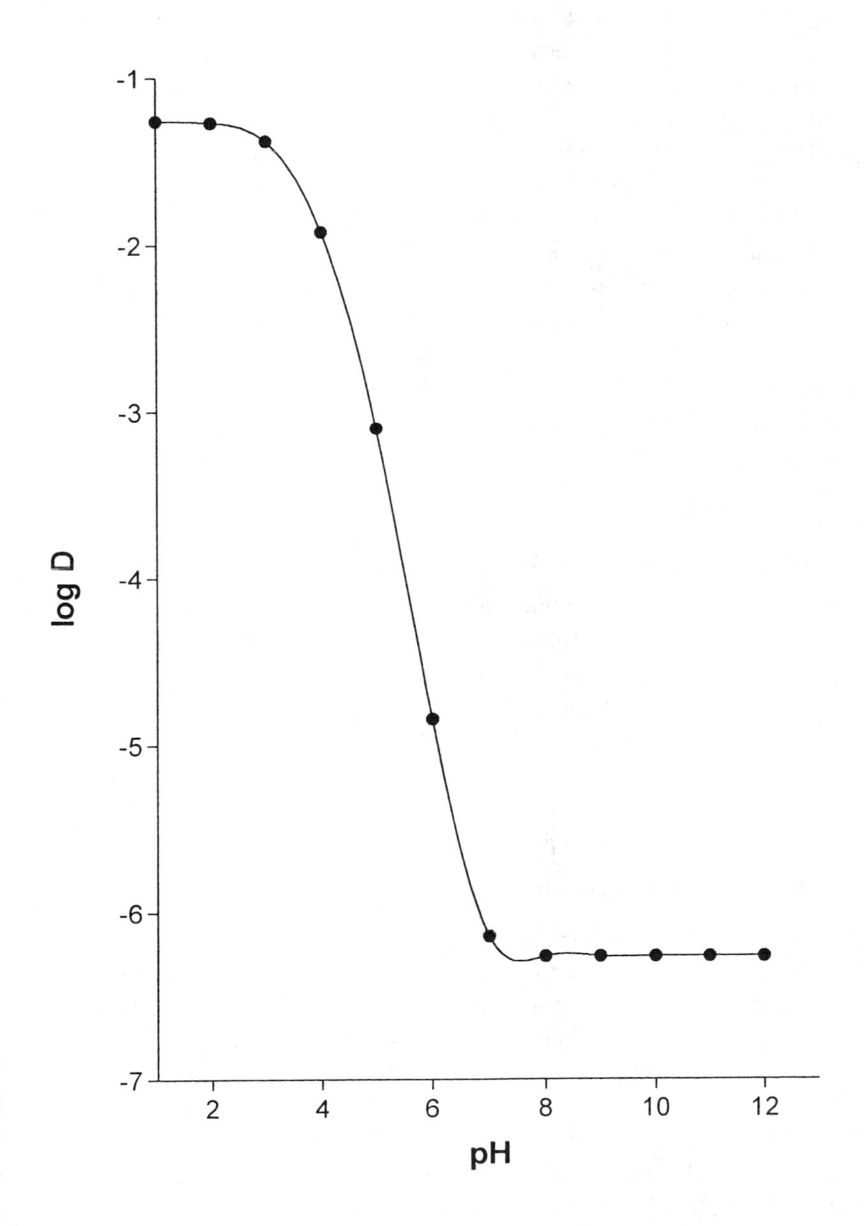

3.5.2 Chiroptical Spectroscopy

Owing to its lack of chromophores, the optical rotatory dispersion (ORD) of malic acid is simple, and it does not exhibit anomalous dispersion [12]. Derivatization with ethyl dithiocarbonate yields a compound having an absorption band around 350 nm, and a consequential anomalous rotatory dispersion curve in that region [13].

Very extensive investigations into the complexation of lanthanide ions by malic acid have been performed using circularly polarized luminescence as a means to study the coordination chemistry of Tb(III) and Eu(III) complexes [14-16]. The metal-ligand binding modes were found to be strongly dependent upon solution pH and details of the complexes formed. For example, Figure 3 illustrates the change in bonding mode that accompanies relatively small changes in pH for the mixed Tb(III) complex with nitrilo-triacetic acid and (*L*)-malic acid.

3.6 Particle Morphology

The particle morphology of malic acid was studied using polarizing optical microscopy. Representative photomicrographs of (*L*)-malic acid and (*DL*)-malic acid are shown in Figures 4 and 5, respectively. Both substances were obtained in the form of aggregated microcrystals, and each exhibits primarily first-order birefringence.

3.7 Crystallographic Properties

3.7.1 Single Crystal Structure

The crystal structure of (*DL*)-malic acid has been reported [17], with the compound crystallizing in the monoclinic $P2_1/n$ space group. The unit cell parameters were reported to be a = 16.865 Å, b = 11.460 Å, c = 12.436 Å, and β = 99.10°. The unit cell was characterized by Z = 4, and a calculated density of 1.586.

In the crystal, (*DL*)-malic acid exists in the *trans* conformation, with the acid forming linear hydrogen bonded chains. The cohesion between the chains was judged to be weak, based only on Van der Waals forces.

Figure 3. Circularly polarized luminescence spectra obtained as a function of pH within the $^5D_4 \rightarrow {}^7F_5$ band system for the mixed Tb(III) complex with nitrilotriacetic acid and (*L*)-malic acid.

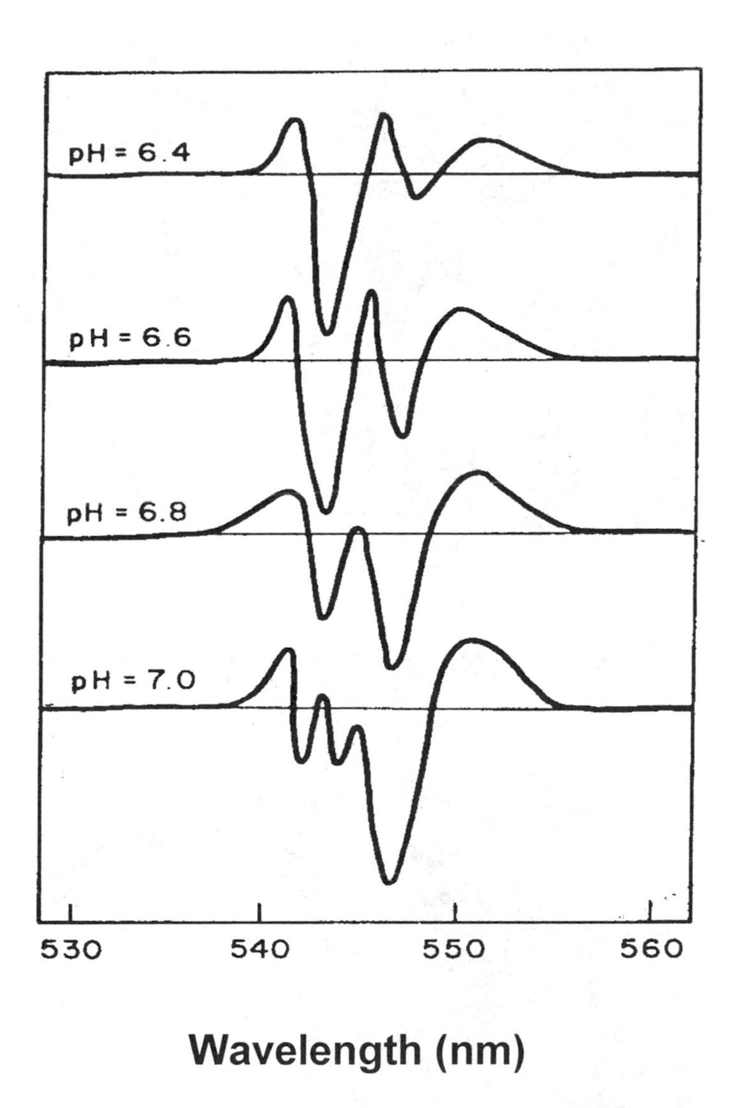

Figure 4. Morphology of (*L*)-malic acid, obtained using light microscopy at a magnification of 100x.

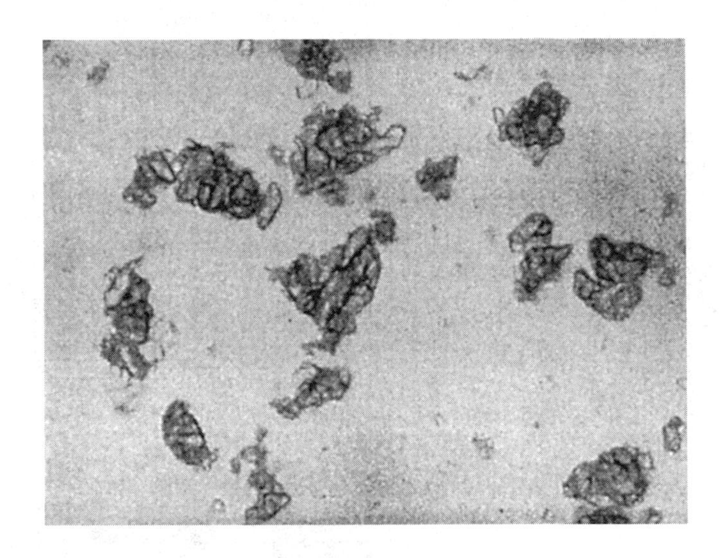

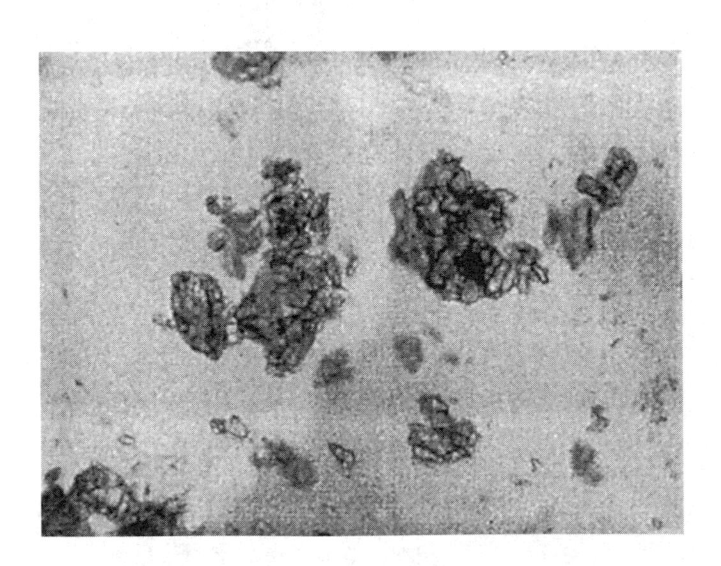

Figure 5. Morphology of (*DL*)-malic acid, obtained using light microscopy at a magnification of 100x.

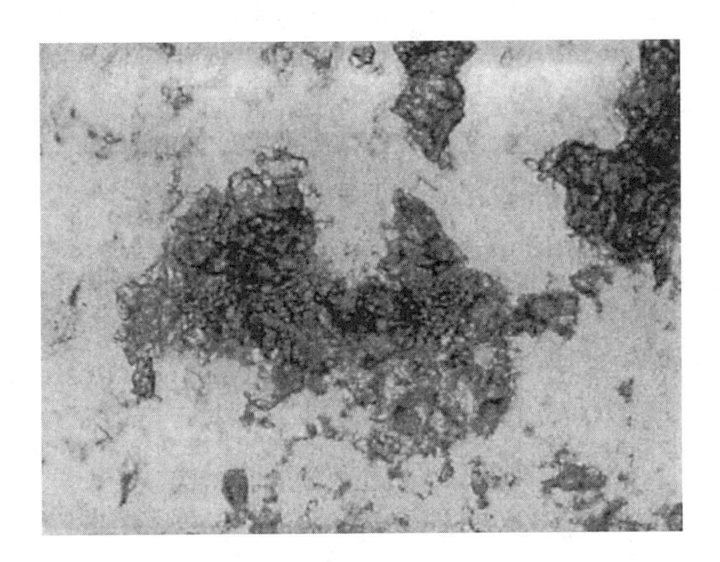

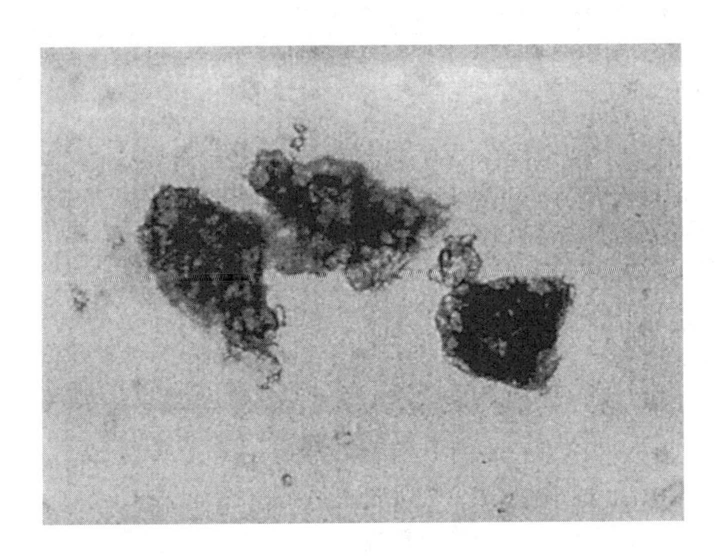

3.7.2 X-Ray Powder Diffraction Pattern

X-ray powder diffraction patterns were obtained for both (*DL*)-malic acid and (*L*)-malic acid using a Philips 7200 system. Since (*DL*)-malic acid has been shown to crystallize in the centrosymmetric $P2_1/n$ space group, and since (*L*)-malic acid must crystallize in a non-centrosymmetric space group, it is not surprising that the powder patterns shown in Figures 6 and 7 are quite different. Tables of crystallographic data obtained from the powder patterns are shown in Tables 3 and 4.

3.9 Thermal Methods of analysis

3.9.1 Melting Behavior

(*DL*)-malic acid has been reported to exhibit a melting point range of 131-132°C, while the melting point range of (*L*)-malic or (*D*)-malic acid has been reported to be 100-101°C [3]. The enantiomerically resolved forms have also been reported to undergo decomposition when heated to approximately 140°C.

3.9.2 Differential Scanning Calorimetry

Differential scanning calorimetry was obtained on (*L*)-malic acid using a TA Instruments model 9020. The thermogram is shown in Figure 8, and consists of a single endotherm identified as the melting phase transition. The onset of melting was noted at 103.3°C, and the peak maximum was observed at 105.3°C. The enthalpy of fusion computed for this sample was calculated as 201 J/g, and the quality of the DSC thermogram suggests that purity determination by DSC would be a viable method.

Differential scanning calorimetry was also obtained on (*DL*)-malic acid, and that thermogram is shown in Figure 9. The sole thermal event observed was a single endotherm identified as the melting phase transition. The onset of melting was observed at 130.8°C, and the peak maximum was noted at 133.8°C. The enthalpy of fusion computed for this sample was calculated as 257 J/g, and the quality of the DSC thermogram suggests that purity determination by DSC would be a viable method.

Figure 6. X-ray powder diffraction pattern of (*L*)-malic acid.

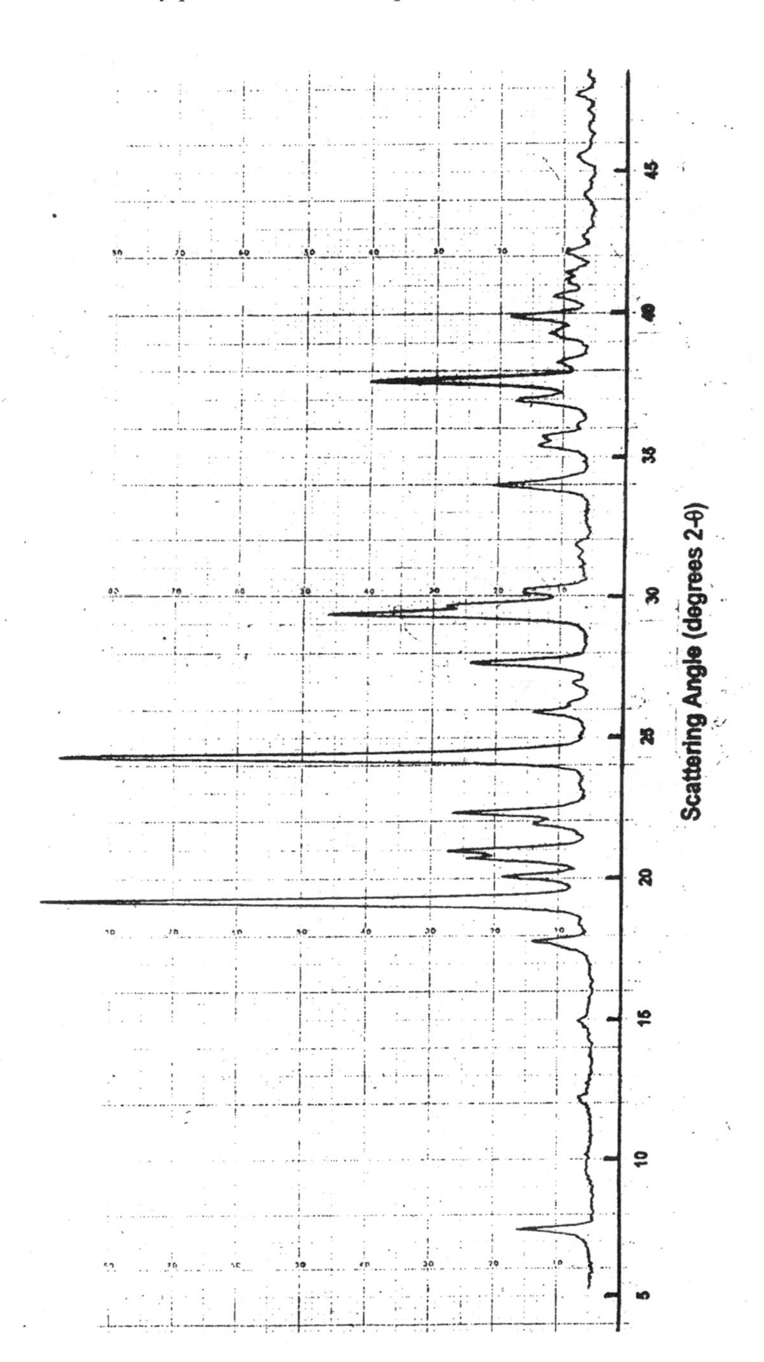

Table 3

Crystallographic Parameters Derived from the X-Ray Powder Pattern of (*L*)-Malic Acid

ID	Scattering Angle (degrees 2-θ)	d-spacing (Å)	Relative Intensity (%)
1	7.4491	11.8846	13.73
2	12.2361	7.2438	1.96
3	14.9635	5.9291	1.96
4	17.8023	4.9895	10.46
5	19.3274	4.5991	100.00
6	20.1401	4.4153	15.69
7	20.7524	4.2864	22.88
8	21.0307	4.2303	25.49
9	21.9547	4.0543	9.80
10	22.3109	3.9904	24.18
11	23.8369	3.7383	1.31
12	24.4261	3.6494	96.08
13	25.9290	3.4412	9.80
14	26.2630	3.3982	3.27
15	26.9866	3.3087	1.96
16	27.7102	3.2239	20.92
17	29.3800	3.0444	46.41
18	29.7140	3.0109	25.49
19	30.1593	2.9675	11.11
20	30.8829	2.8996	1.31
21	31.3839	2.8544	1.96
22	31.8292	2.8155	2.61

Table 3 (continued)

Crystallographic Parameters Derived from the X-Ray Powder Pattern of (*L*)-Malic Acid

ID	Scattering Angle (degrees 2-θ)	d-spacing (Å)	Relative Intensity (%)
23	34.0000	2.6406	16.34
24	35.4472	2.5360	8.50
26	35.7812	2.5131	7.84
27	37.8964	2.3776	1.96
28	38.7869	2.3250	13.07
29	39.4549	2.2872	39.87
30	40.1228	2.2506	5.23
31	41.1248	2.1981	7.19
32	41.7927	2.1645	13.73
33	42.4050	2.1346	6.54
34	42.9060	2.1109	4.58
35	43.2399	2.0953	4.58
36	43.9635	2.0625	5.23
37	44.4088	2.0429	1.31
38	45.1881	2.0094	1.31
39	46.0230	1.9749	1.31
40	46.5240	1.9548	0.65

Figure 7. X-ray powder diffraction pattern of (*DL*)-malic acid.

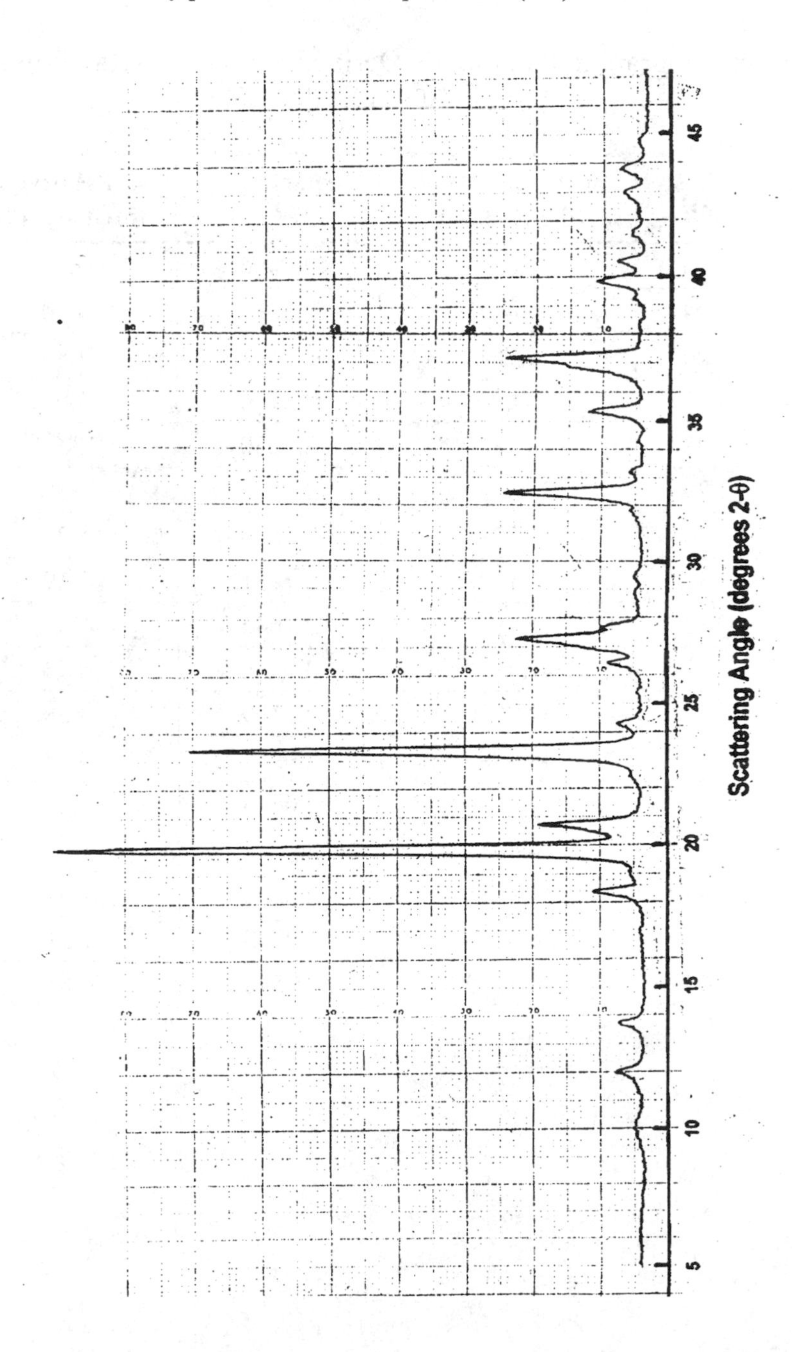

Table 4

Crystallographic Parameters Derived from the X-Ray Powder Pattern of (*DL*)-Malic Acid

ID	Scattering Angle (degrees 2-θ)	d-spacing (Å)	Relative Intensity (%)
1	10.1197	8.7535	1.84
2	12.0675	7.3447	4.29
3	13.7926	6.4297	4.91
4	18.4115	4.8258	7.98
5	19.9140	4.4649	100.00
6	20.7487	4.2872	16.56
7	23.4199	3.8039	73.01
8	24.3659	3.6583	4.29
9	25.4789	3.5010	1.23
10	26.4806	3.3708	5.52
11	27.3153	3.2696	20.25
12	27.7049	3.2245	6.13
13	28.8179	3.1025	1.23
14	29.4300	3.0393	1.84
15	31.8229	2.8161	2.45
16	32.3794	2.7689	22.09
17	44.3440	2.0457	1.84
18	34.2715	2.6203	0.61
19	35.3845	2.5404	8.59

Table 4

Crystallographic Parameters Derived from the X-Ray Powder Pattern of (*DL*)-Malic Acid

ID	Scattering Angle (degrees 2-θ)	d-spacing (Å)	Relative Intensity (%)
20	36.8870	2.4403	12.27
21	37.2209	2.4191	22.09
22	38.3339	2.3514	0.61
23	39.2243	2.3001	1.84
24	39.8364	2.2661	7.36
26	40.5599	2.2274	4.29
27	41.5616	2.1760	2.45
28	43.0084	2.1061	3.07
29	43.8432	2.0679	3.68
30	44.7892	2.0264	1.23

Figure 8 Differential scanning calorimetry thermogram of (*L*)-malic acid.

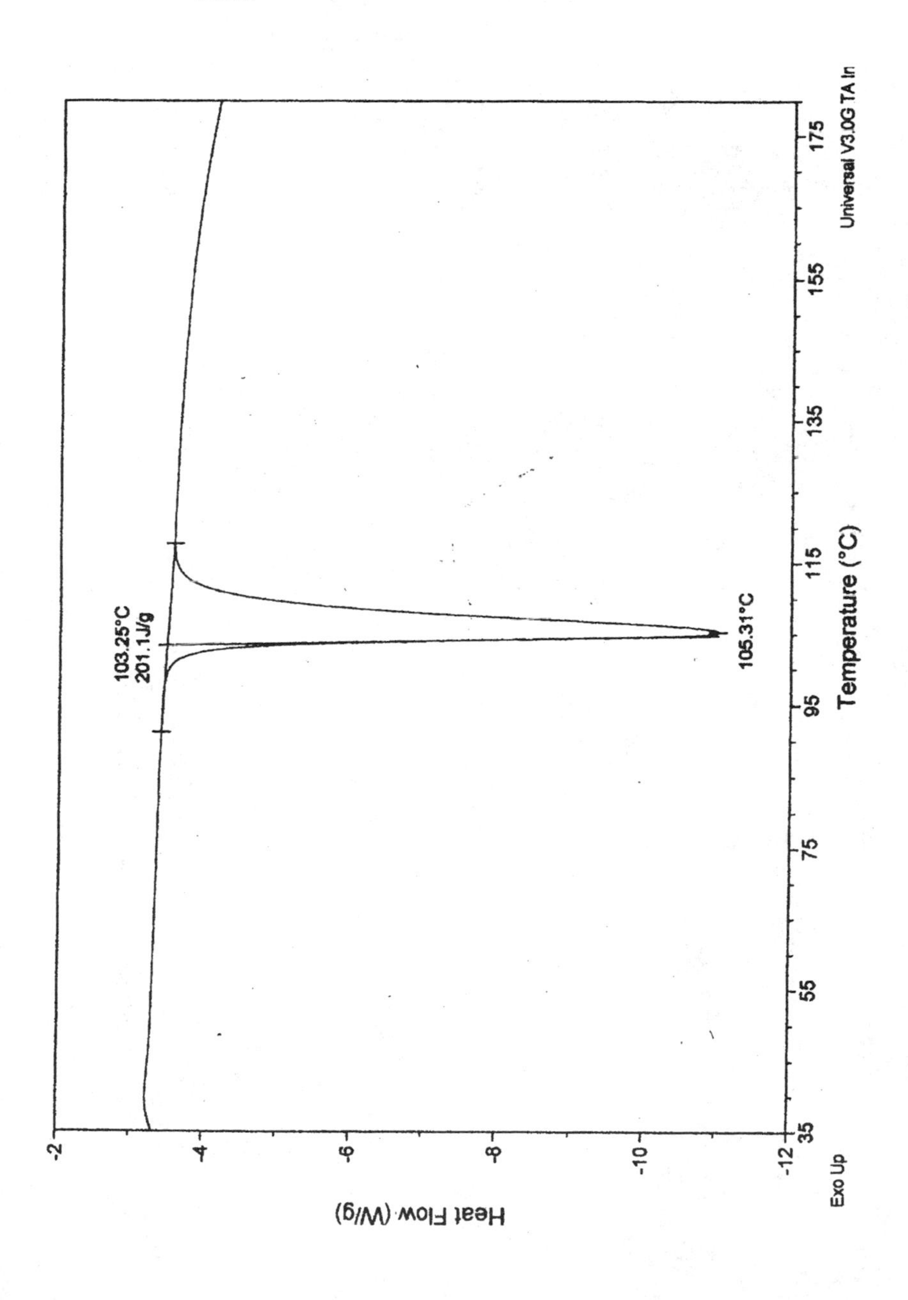

Figure 9 Differential scanning calorimetry thermogram of (*DL*)-malic acid.

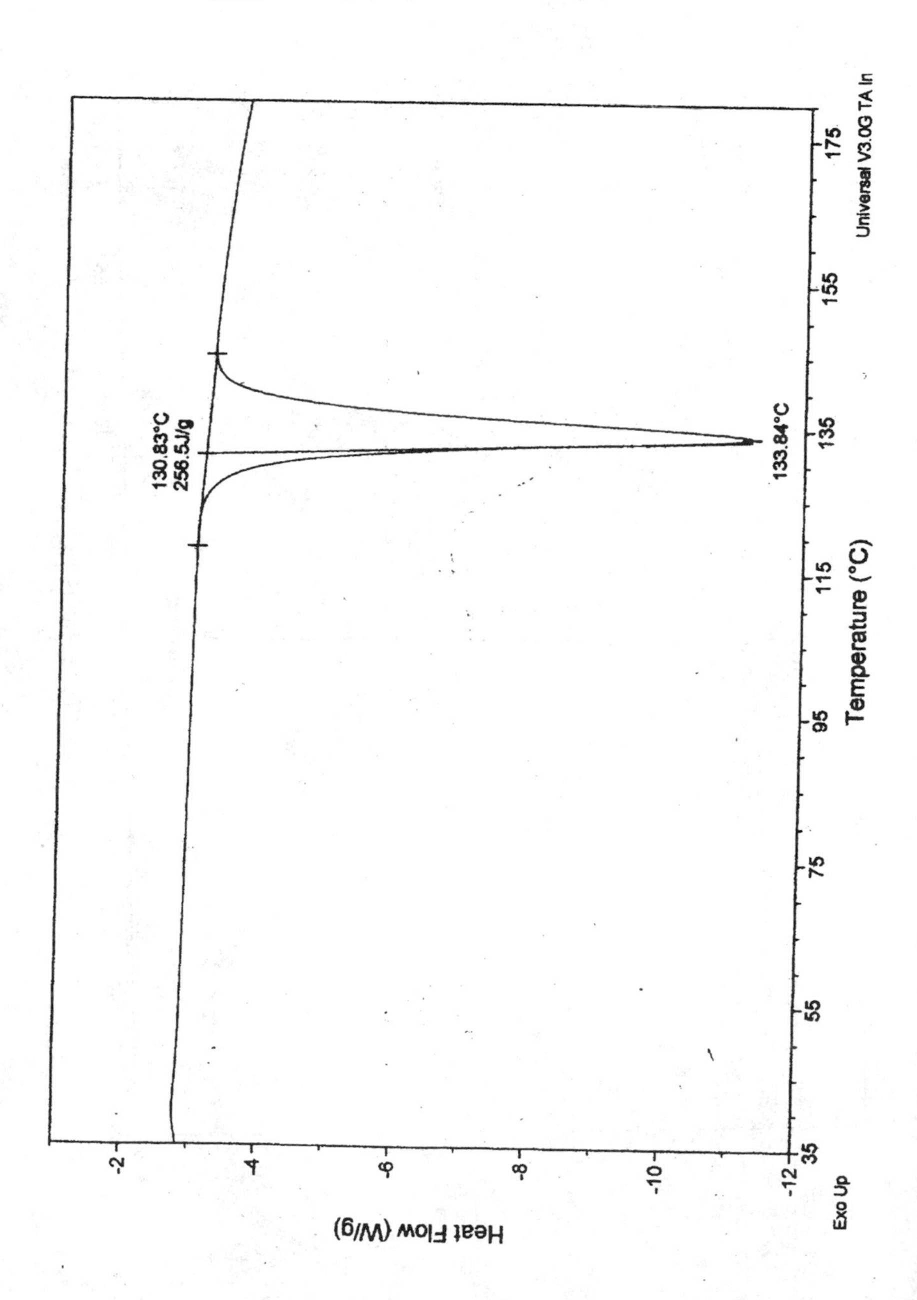

3.10 Spectroscopy

3.10.1 UV/VIS Spectroscopy

Malic acid does not possess any chromophores that yield absorbance in the visible or ultraviolet regions of the spectrum. However, a residual end-absorption is noted, with this resulting from the long-wavelength tail of the carboxylate absorption band. Hence, chromatographic methods based on the solute absorbance at 200-210 nm are possible.

3.10.2 Vibrational Spectroscopy

As a relatively simple molecule, malic acid exhibits an uncomplicated pattern of vibrational modes. Infrared absorption spectra for both (*L*)-malic acid (Figure 10) and (*DL*)-malic acid (Figure 11) have been reported in standard compilations [18].

The infrared absorption and Raman spectra of (*DL*)-malic acid have been contrasted [19], and that comparison is shown in Figure 12. The most intense bands were listed in the published work, and these are listed in Table 5. The assignments for these vibrational modes have been made by this author [8], and these are also found in Table 5.

3.10.3 Nuclear Magnetic Resonance Spectrometry

3.10.3.1 ^{1}H-NMR Spectrum

The ^{1}H-NMR spectrum of (*L*)-malic acid has been reported in a standard compilation [20], and is shown in Figure 13. Assignments for the various resonance bands are found in Table 6.

3.10.3.2 ^{13}C-NMR Spectrum

The solid-state ^{13}C-NMR spectrum of (*DL*)-malic acid has been reported [19], and is shown in Figure 14. The carbons of the two carboxylate groups are observed at 181.3 ppm, the carbon bearing the hydroxyl group is observed at 68.5 ppm, and the methylene carbon is observed at 40.6 ppm.

Figure 10 Infrared absorption spectrum of (*L*)-malic acid [18].

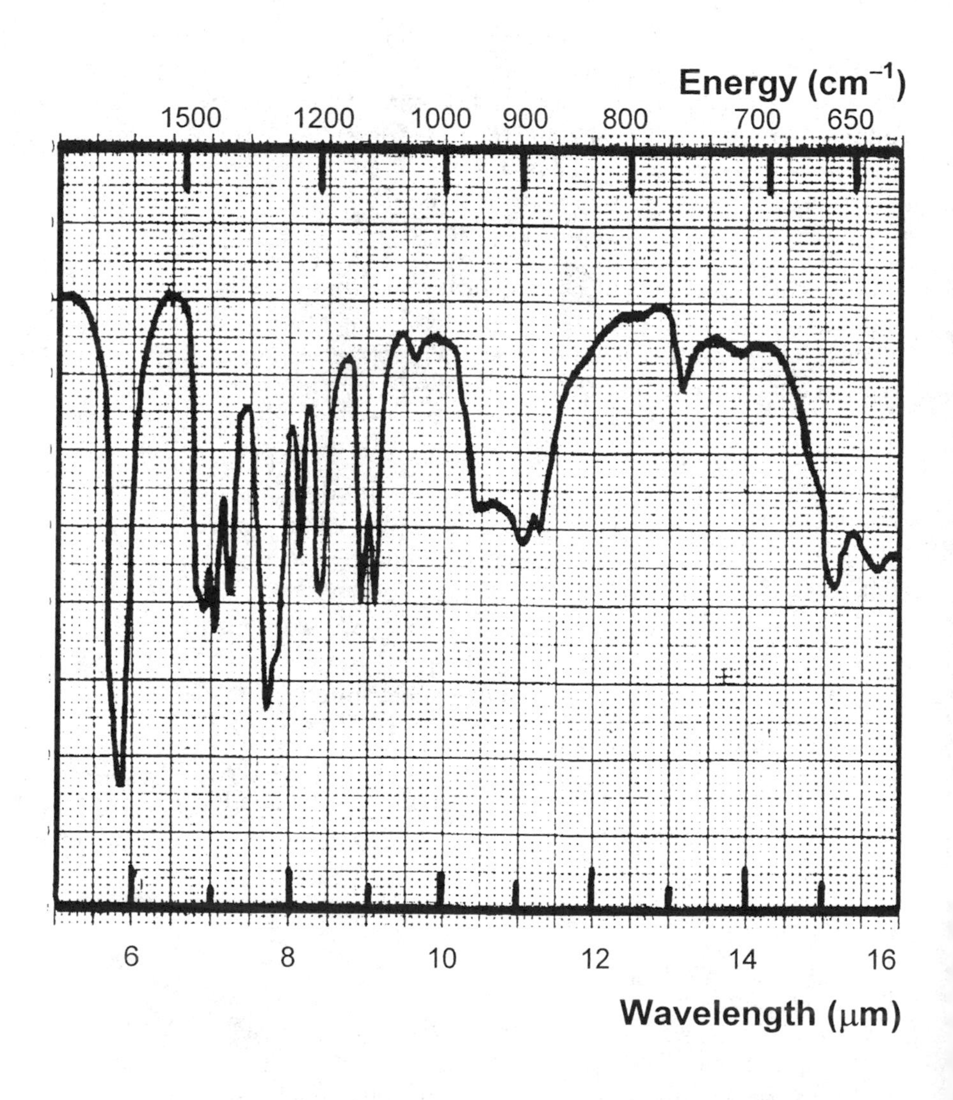

Figure 11 Infrared absorption spectrum of (*DL*)-malic acid [18].

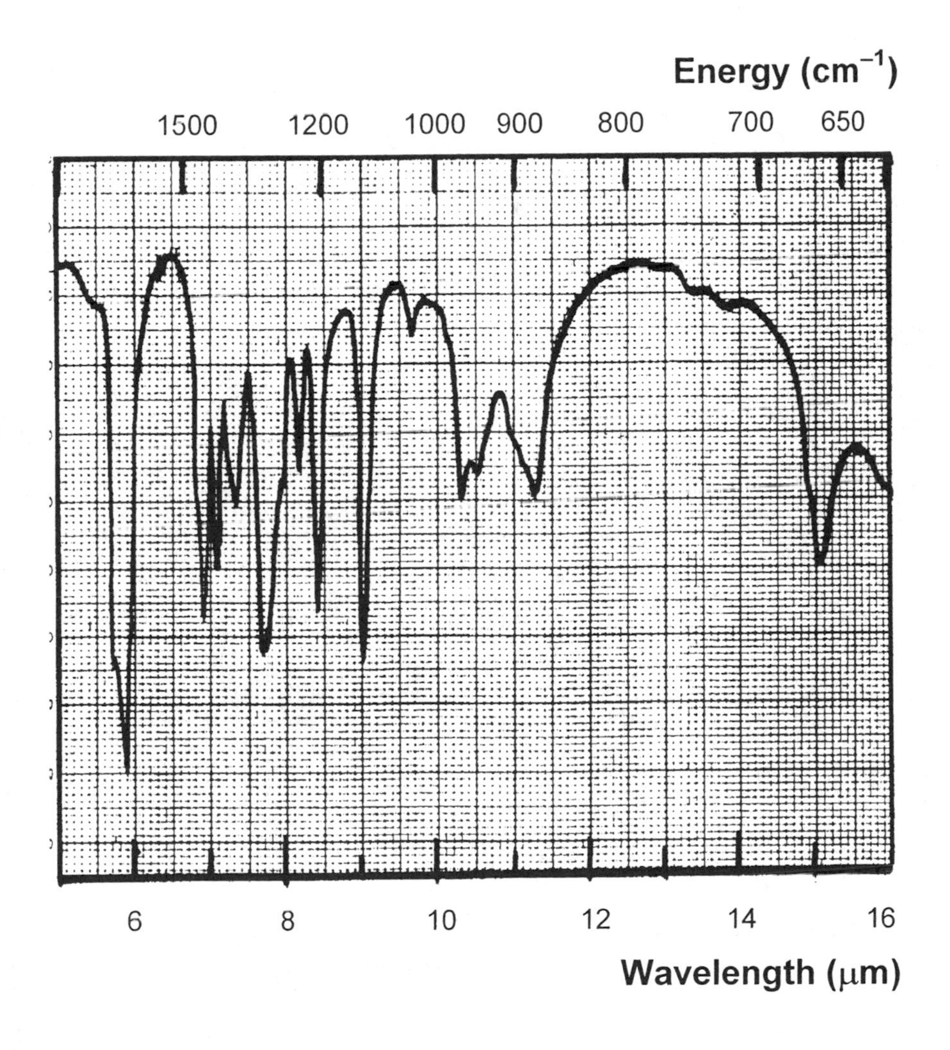

Figure 12 Infrared absorption (upper trace) and Raman (lower trace) spectra of (*DL*)-malic acid [19].

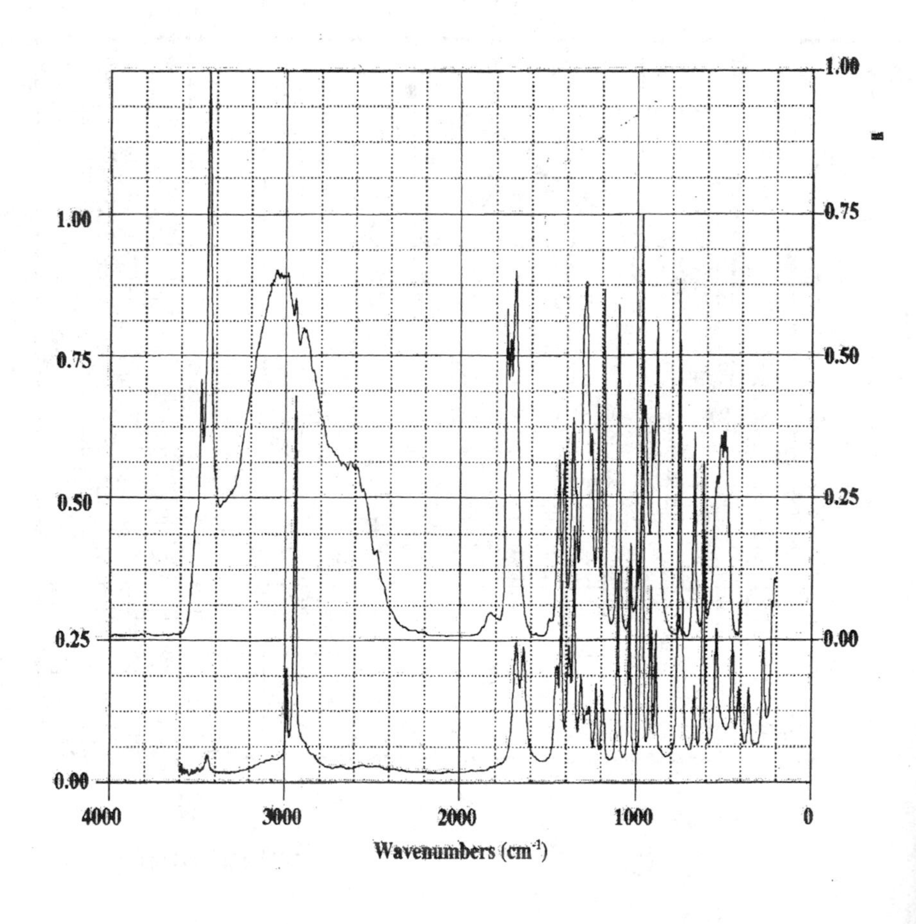

Table 5

Assignments for the Observed Vibrational Modes of (*DL*)-Malic Acid

Infrared Absorption (cm^{-1})	**Raman Difference (cm^{-1})**	**Assignment**
1738		Asymmetric C=O stretch (monomer)
1716		Asymmetric C=O stretch (dimer)
1689		Symmetric C=O stretch
	1422	methylene deformation
	1349	–OH deformation in plane
1288		C–O stretch (dimer)
1185		
1103	1097	CH_2-CHOH-CH_2 stretch
	1032	
968	963	–OH deformation out of plane (dimer)
885		
	911	
	748	– CH_2 rocking mode
	610	–OH deformation (acid)
	532	–OH deformation (hydroxy)

Figure 13 ^{1}H-NMR spectrum of (*L*)-malic acid [20].

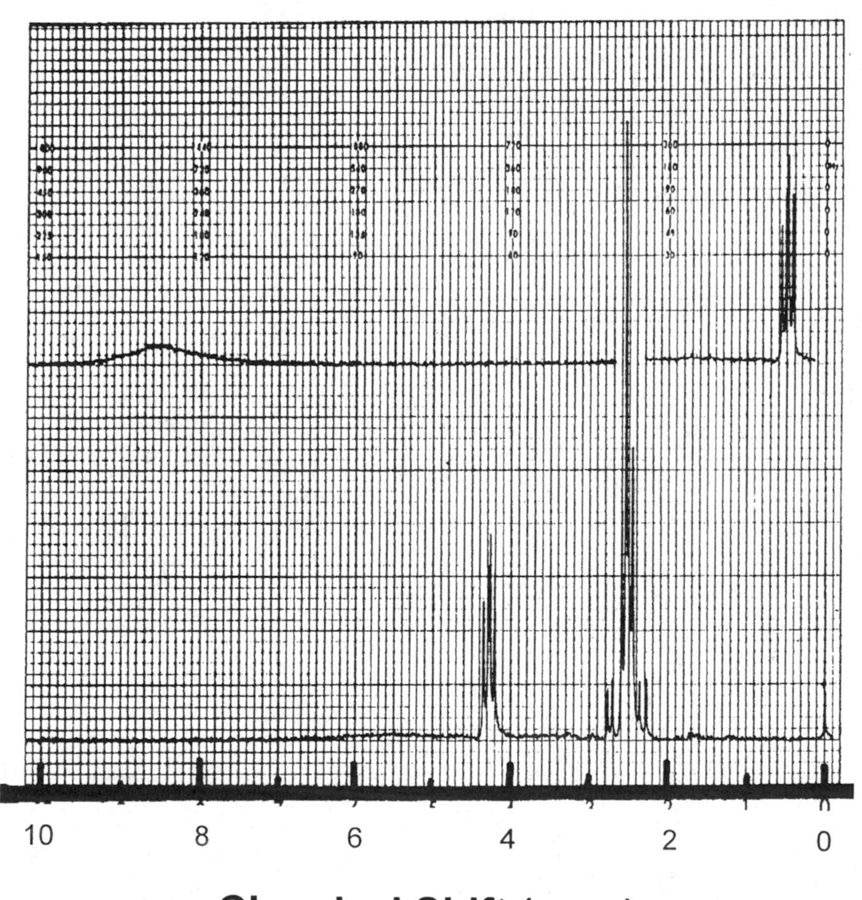

Table 6

Assignments for the Observed Resonance Bands in the ^{1}H-NMR Spectrum of (*L*)-Malic Acid

```
         b   e
    O  H  OH  O
 d  ||  |  |  ||   d
HO-C -C -CH -C -OH
       |   c
       H
       a
```

Hydrogen	Chemical Shift (ppm)	Multiplicity
H-**a**	2.43	1
H-**b**	2.53	1
H-**c**	4.28	1
H-**d**	12.36	2
H-**e**	undetected	1

Figure 14 ^{13}C-NMR spectrum of (*DL*)-malic acid [19].

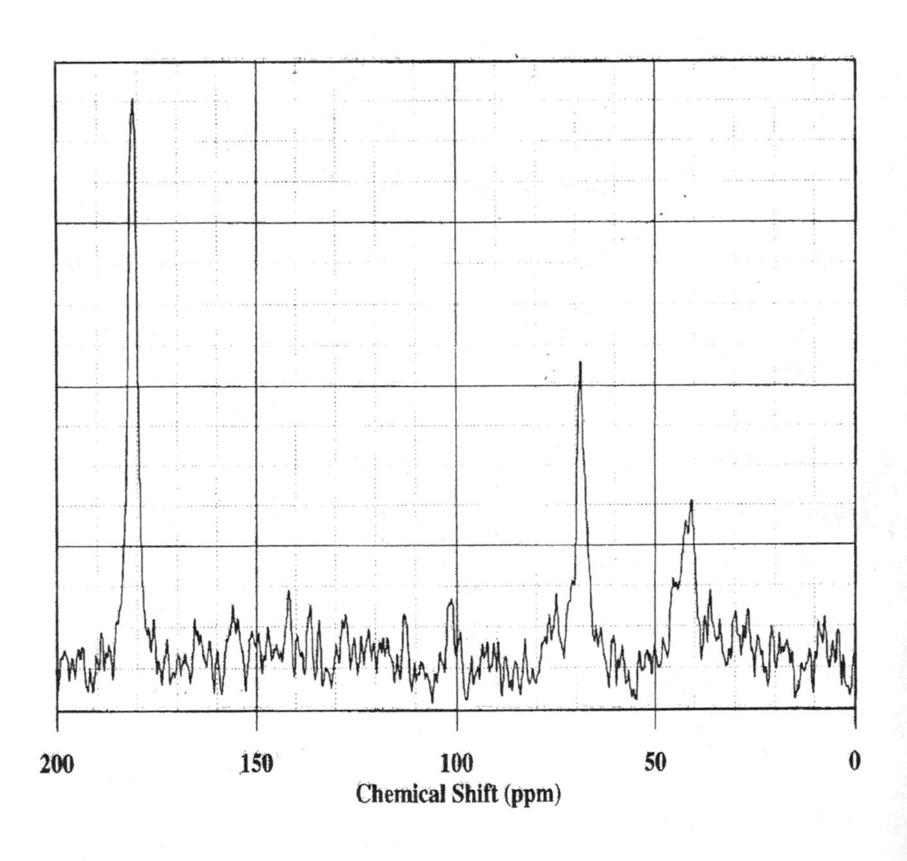

4. Methods of Analysis

4.1 USP 24 Compendial Tests [21]

Malic Acid contains not less than 99.0 percent and not more than 100.5 percent of $C_4H_6O_5$. A USP reference standard is available.

4.1.1 Identification

The identification is verified on the basis of characteristic infrared absorption, following the directives of general test <197K>, which is conducted on an undried specimen. The IR spectrum of the sample must be equivalent to that of the reference standard processed in the same manner.

4.1.2 Residue on Ignition

When performed according to the procedure of general test <281>, acceptable material yields not more than 0.1%.

4.1.3 Water Insoluble Substances

This quantity is determined using the following procedure. Dissolve 25 g of malic acid in 100 mL of water, filter the solution through a tared filtering crucible, and wash the filter with hot water. The crucible and contents are dried at 100°C to constant weight. The amount of mass increase in the crucible is not more than 25 mg (0.1%).

4.1.4 Heavy Metals

When performed according to the procedure of general test <231>, Method II, the substance does not contain more than 0.002% recoverable sulfide precipitate.

4.1.5 Fumaric and Maleic Acid Content

Any fumaric or maleic acid impurities in a sample of malic acid are determined using high pressure liquid chromatography. In this procedure, the mobile phase is filtered and degassed 0.01 N sulfuric acid in water.

The *Standard Solution* is prepared using the mobile phase as a solvent, and is an accurately known concentration around 0.005 mg/mL of fumaric acid RS and about 0.002 mg/mL of maleic acid RS. The *Test Preparation* is prepared by transferring about 100 mg, accurately weighed, of malic Acid to a 100-mL volumetric flask, and then dissolving in and diluting with mobile phase to volume. The *Resolution Solution* is also prepared using mobile phase as the solvent, and consists of a solution containing about 1 mg/mL of malic acid, about 10 μg/mL of fumaric acid RS, and about 4 μg/mL of maleic acid RS.

The liquid chromatograph is equipped with a 210-nm detector and a 6.5-mm x 30-cm column that contains packing L17 (strong cation-exchange resin consisting of sulfonated cross-linked styrene-divinylbenzene copolymer in the hydrogen form, 7 to 11 μm in diameter). The temperature of the column is maintained at 37 ± 1°C, and the flow rate is about 0.6 mL/minute.

To perform the assay, one chromatographs the *Resolution Solution*, and records the peak responses. An acceptable system is one for which the resolution of the maleic acid and malic acid peaks is not less than 2.5, the resolution of the malic acid and fumaric acid peaks is not less than 7.0, and the relative standard deviation of the maleic acid peak for replicate injections is not more than 2.0%. If the system is found to be suitable, then one separately injects equal volumes (about 20 μL) of the *Standard Preparation* and the *Test Preparation* into the chromatograph, records the chromatograms, and measures the peak responses. The relative retention times are about 0.6 for maleic acid, 1.0 for malic acid, and about 1.5 for fumaric acid. The quantities (in units of mg) of maleic acid and of fumaric acid in the sample taken is calculated using:

$$\text{mg (MA)} = 100\,C\ \{r_U / r_S\}$$

where C is the concentration (in units of mg/mL) of the corresponding reference standard in the *Standard Preparation*, and r_U and r_S are the responses of the corresponding peaks from the *Test Preparation* and the *Standard Preparation*, respectively. The specification is that not more than 1.0% of fumaric acid and not more than 0.05% of maleic acid are found.

4.1.6 Organic Volatile Impurities

When performed according to the procedure of general test <467>, Method I, the tested substance meets the following requirements:

Benzene, NMT 2 μg/g
Chloroform, NMT 60 μg/g
1,4-Dioxane, NMT 380 μg/g
Methylene Chloride, NMT 600 μg/g
Trichloroethylene, NMT 80 μg/g

4.1.7 Assay

The assay value for malic acid is determined using potentiometry. The procedure is conducted by transfer about 2 g of accurately weighed malic acid to a conical flask, dissolving in 40 mL of recently boiled and cooled water, adding phenolphthalein TS, and then titrating with 1 N NaOH VS to the first appearance of a faint pink color that persists for not less than 30 seconds. Each milliliter of 1 N NaOH is equivalent to 67.04 mg of $C_4H_6O_5$.

4.2 AOAC Methods of Analysis [22]

The American Organization of Analytical Chemists has reported Official Methods 968.19 and 933.07 suitable for the determination of (*L*)-malic and (*DL*)-malic acid, respectively, in fruits and fruit products.

4.2.1 AOAC Method 968.19

Application of the method requires access to a polarimeter having accuracy to at least 0.01°, and which is equipped with a sodium vapor lamp operating at 589 nm.

The (*L*)-malic acid *Stock Solution* (10 mg/mL) is prepared by placing 1.0 g of reference standard (*L*)-malic acid and 4.0 g citric acid in a 150-mL beaker. To this is added about 50 mL water, and the pH adjusted to pH 5.5 with 50% NaOH solution. This solution is quantitatively transferred to a 100-mL volumetric flask, and diluted to volume with water. *Working Solutions* at concentrations of 1, 2, and 3 mg/mL are prepared by pipet 10, 20, and 30 mL of the *Stock Solution* to separate 100 mL volumetric flasks and diluting to volume with water. The optical rotations of these solutions is determined in a 200 mm polarimeter tube.

A calibration curve is prepared by placing about 35 mL of each *Working Solution* in 50-mL flasks, and adding 1.5 g uranyl acetate dihydrate. The solutions are kept in the dark for 30 minutes with occasional swirling. After that, the solutions are filtered, and the optical rotation of the clear uranyl malate complex is determined in a 200 mm polarimeter tube.

A number of sample preparative methods are described in the Official Method, but only the method used for fruit juices will be described here. One begins by weighing 62.5 g of juice into a 250-mL volumetric flask, adding 1 g of potassium acetate and 100 mL alcohol, and mixing. The solution is diluted to volume with alcohol, mixed, and allowed to stand for 1 hour. After this period, the solution is filtered through rapid paper.

To run the procedure, transfer 200 mL of the aliquot filtrate to a 250-mL wide-mouth bottle, 5.7 cm od x 13 cm high, and add a magnetic stirring bar and 30 mL absolute alcohol. Potentiometrically titrate 10 mL of the remaining filtrate to pH 8.4, using 0.1N NaOH, and calculate the volume of NaOH required to neutralize the 200 mL aliquot. Add 0.6 mL of saturated lead acetate solution to the bottle for each milliliter of 1N NaOH calculated to neutralize the 200 mL aliquot. Stir 10 minutes, and centrifuge for 6 minutes at 1500 rpm. Test the supernatant for complete precipitation with few drops saturated lead acetate solution. Decant and wash the precipitate by stirring 5 minutes with 200 mL alcohol. Centrifuge for 5 minutes, decant add 25 mL water to the precipitate, and mix well to form a slurry. Using a pH meter, adjust the pH to 1.5 with 10% sulfuric acid. Remove the lead sulfate precipitate by vacuum filtration, using a coarse porosity 60-mL fritted glass crucible containing an asbestos pad. Wash the precipitate with 10 mL portions of water, and combine the washings with the filtrate in a 150-mL graduated beaker. The total volume should be less than 90 mL. Adjust the pH to 5.5 with 50% NaOH, transfer quantitatively to a 100-mL volumetric flask, and dilute to volume with water. Add approximately 6 g of activated charcoal, and mix thoroughly. Let the solution stand for 30 minutes, and filter through fine filter paper (the filtrate must be colorless). Determine the optical rotation of this filtrate in a 200-mm polarimeter tube. To approximately 35 mL of the filtrate in a 50-mL flask, add 1.5 g uranyl acetate dihydrate, and keep in the dark for 30 minutes with occasional swirling. Filter and determine the optical rotation of the clear solution of uranyl malate complex.

The number of milligrams of (*L*)-malic acid per 100 g sample is calculated using:

$$\text{mg}\ (L\text{-mal}) = \{ (\alpha_x)\ (C)\ (100) \} \ / \ \{ (\alpha_S)\ (W) \}$$

where (α_x) is the difference in optical rotation of the Sample with and without the uranyl ion, (α_S) is the difference in optical rotation of the Standard with and without the uranyl ion, C is the number of mg of (L)-malic acid in 100 mL of the Standard Solution, and W is the number of grams of sample in 100 mL of the final solution.

Alternatively, the number of milligrams of (*L*)-malic acid per 100 g sample can be calculated as (α_x) multiplied by twice the sum of mg of (*L*)-malic acid in the standard curve solutions divided by the sum of the number degrees in the standard curve solutions. Finally, the same result can be calculated as twice the number of mg of (*L*)-malic acid from the standard curve.

4.2.1 AOAC Method 933.07

The AOAC specifies that the method is empirical, requiring that all directions be rigidly followed, particularly with respect to dilutions. Substitution of volumetric flasks of capacities different from those specified is not permissible.

Tribasic lead acetate is prepared by dissolving 82 g of $Pb(CH_3COO)_2{\cdot}3H_2O$ in 170 mL water. Prepare 100 mL dilute NH_4OH solution containing 5.8 g NH_3 as determined by titration (methyl red as indicator). Heat solutions to 60°C, mix thoroughly, and let stand overnight. Shake vigorously to break up precipitate, and filter on a Büchner. Wash once with water and vacuum dry, then wash twice with alcohol, and finally with ether. Let the solid dry in air.

A number of reagent solutions are required for this procedure. A *Lead Acetate Solution* is prepared by dissolving 40 g $Pb(CH_3COO)_2{\cdot}3H_2O$ in water, adding 0.5 mL CH_3COOH, and diluting to 100 mL. *Tribasic Lead Acetate Standard Solution* is prepared by placing 5 g of the salt in a 500-mL Erlenmeyer flaks, adding 200 mL of water, and shaking vigorously.

Neutralize 3 mL 1N sulfuric acid (diluted with 200 mL water) with the solution using methyl red as an indicator. Note the volume of lead solution required, and in determination use 2 mL in excess of this volume. Prepare an *Oxalic Acid Standard Solution* by dissolving 28.7556 g purest $H_2C_2O_4 \cdot 2H_2O$ in water, and diluting to 1 liter. Finally, prepare a *Potassium Permanganate Standard Solution* by dissolving 14.5214 g the purest $KMnO_4$ in water, and diluting to 1 liter. The $KMnO_4$ solution is to be standardized using the *Oxalic Acid Standard Solution*, following this procedure. Pipet 50 mL the *Oxalic Acid Standard Solution* into a 600-mL beaker, and add 70 mL water and 10 mL 50% H_2SO_4. Heat to 80°C, immediately add KMn04 solution to a faint pink color, again heat to 80°C, and finish the titration. 50 mL of the *Potassium Permanganate Standard Solution* is equivalent to 50 mL of oxalic acid solution.

Divide the sample under analysis into two portions, one for determination of (*L*)-malic acid by polarimetry, and the other for total malic acid. Choose an amount of sample that will yield a titratable acidity that is less than 150 mg of acid calculated as malic acid. Designate as X mL the quantity of 1N NaOH required to neutralize amount of sample chosen. Adjust the sample solution to approximately Ca 35 mL by evaporation or by addition of water, pour into a 250-mL volumetric flask. Rinse with 10 mL hot water, and then with alcohol, and finally dilute to volume with alcohol. Shake and let stand until the pectin separates, leaving a clear liquid (overnight if necessary), and filter through folded paper. Drain thoroughly and cover the funnel with watch glass. Pipet 225 mL of the filtrate into a centrifuge bottle.

The method requires a prior isolation of total malic acid prior to its determination. To the solution in the centrifuge bottle, add approximately 25 mg of citric acid and a volume of *Lead Acetate Solution* equal to X (X + 3 mL if saponification was performed), shake vigorously for 2 minutes, and centrifuge. Carefully decant the supernatant from precipitated lead salts and test with a small amount of *Lead Acetate Solution.* If a precipitate forms, return to the centrifuge bottle, add more *Lead Acetate Solution*, shake, and centrifuge again. If any sediment lifts, repeat the centrifuging, increasing both speed and time. Let the precipitate drain thoroughly by inverting the bottle several minutes. Add 200 mL of 80% alcohol, shake vigorously, and again centrifuge, decant, and drain. To lead salts add about 150 mL of water, shake vigorously, and pass in a rapid stream of H_2S to saturation. Stopper the bottle and shake for about 1 minute. Transfer the mixture to a 250-mL volumetric flask with water, dilute to volume, shake, and filter through folded paper. Pipet 225 mL of the filtrate into a 600 mL beaker, and evaporate to about 100 mL to expel H_2S. Transfer to a 250-mL volumetric flask with water, and the volume in the flask should be around 200 mL. Add 5 mL 10% acetic acid and the same amount of *Lead Acetate Solution* previously used. Shake vigorously, dilute to volume with water, and filter. Pass a rapid stream of H_2S into the clear filtrate to saturation, stopper the flask, shake vigorously, and filter. Pipet 225 mL of the filtrate into a 600-mL beaker, add about 75 mg tartaric acid, and evaporate over burner and gauze to around 50 mL. Cool, neutralize with 1N KOH (phenolphthalein endpoint), and add 5 drops in excess. Add 2 mL acetic acid, and transfer the mixture to a 250-mL volumetric flask with alcohol. Dilute to volume with alcohol, shake, and pour into a 500-mL Erlenmeyer flask. Add a small handful of glass beads and cool to 15°C. Stopper the flask, shake vigorously for 10 minutes, and place in a refrigerator for 30 minutes. Again shake 10 minutes, and filter through folded paper. Adjust the clear filtrate to 20°C and pipet 225 mL into a centrifuge bottle. Add *Lead Acetate Solution* equal to X (X + 3 mL if saponification was performed), shake vigorously for about 2 minutes, centrifuge, decant, and drain. Add 200 mL of 80% alcohol, shake, centrifuge, decant, and drain. Transfer the lead salts to a 500 mL Erlenmeyer flask with about 175 mL of water. Add 3 mL of 1N H_2SO_4 and heat to boiling, and then add 1 mL of 5% CH_3COOH and the volume of *Tribasic Lead Acetate Standard Solution* previously determined. Boil the mixture for 5 minutes, cool to room temperature, transfer to a 250-mL volumetric flask with water, dilute to volume, shake, and pour into a 500 mL Erlenmeyer flask. Add a small handful of glass beads, cool to about 15°C, shake vigorously 5 minutes, and place in refrigerator for 30 minutes. Again shake for 5 minutes, and filter through folded paper. Saturate the clear filtrate with H_2S, shake vigorously, and filter. Use one of the two portions for the polarimetry phase and the other for the oxidation phase of the method.

Polarimetry is used to determine the amount of (*L*)-malic acid in the sample. Evaporate 225 mL of the clear filtrate over a burner and gauze to about 10 mL, neutralize with 1N KOH (phenolphthalein endpoint), make slightly acid with 5% CH_3COOH, and evaporate to about 5 mL. Transfer to a 25-27.5 mL Giles flask with water, dilute to the 27.5 mL mark, shake, and pour into a small glass-stoppered flask. If a Giles flask is not available, use a 25 mL graduated cylinder, dilute to volume, and add 2.5

mL of water from a buret. Add a small handful of glass beads and 4 g powdered uranyl acetate, shake vigorously for 10 minutes, and filter. Since the uranyl malate complex is light sensitive, wrap the flask in a towel while shaking and protect from light as much as possible during filtration and polarimetry. Measure the optical rotation in a 200 mm polarimetry tube at 20°C. The amount of (L)-malic acid in the sample is determined as described in the preceding section for AOAC Official Method 968.19.

An oxidative titration is used to determine the total amount of malic acid present in the sample. Evaporate 225 mL of the clear sample solution to about 10 mL to expel all alcohol, dilute to about 120 mL with water, and add 10 mL 30% NaOH solution and 25 mL of *Potassium Permanganate Standard Solution.* Heat to approximately 80°C and keep in a boiling water bath for 30 minutes. Add 25 mL of *Oxalic Acid Standard Solution* and 10 mL of 50% H_2SO_4, stirring vigorously. Adjust to 80°C, and titrate to faint pink with *Potassium Permanganate Standard Solution.* Again heat to 80°C and finish the titration. Each milliliter of *Potassium Permanganate Standard Solution* used multiplied by 5 equals the total oxidizable material (as malic acid) present in aliquot.

Calculate the number of milligrams of (*DL*)-malic acid in the sample taken for analysis using:

$$\text{mg}\ (DL\text{-mal}) = 4\,(\,T - 5 - L\,)$$

where T = mg oxidizable as malic acid, L = mg (*L*)-malic acid, 5 is a correction factor for the number of mg of non-malic material measured as malic acid, and 4 is a factor correcting for the reversion of inactive malic acid in the aliquot back to an amount of inactive acid in the sample taken for analysis.

5. Stability [1]

Malic acid is stable at temperatures up to 150°C, and at temperatures above 150°C it loses water very slowly to yield fumaric acid. Complete decomposition occurs at about 180°C, yielding fumaric acid and maleic anhydride. The bulk material should be stored in a well-closed container, in a cool, dry, place, since conditions of high humidity and elevated temperatures lead to caking

Malic acid is readily degraded by many aerobic and anaerobic microorganisms, and will react with oxidizing materials.

6. Safety [1]

Malic acid is used in a variety of oral, topical, and parenteral formulations, and is regarded as a non-toxic and non-irritant material. Concentrated solutions may pose an irritation threat, however.

The following LD values have been reported:

LD_{50} (mouse, oral) = 1.6 g/kg

LD_{50} (rat, oral) = 4.73 g/kg

The substance may be handled following normal precautions associated with solid organic acids. Concentrated solutions require protection from their irritant properties.

7. References

1. ***Handbook of Pharmaceutical Excipients***, 3rd edn., A.H. Kibbe, ed., American Pharmaceutical Association, Washington, D.C., 2000, pp. 311-312.

2. ***Martindale, The Extra Pharmacopoeia***, 30th edn., J.E.F. Reynolds, ed., Pharmaceutical Press, London, 1993, p. 1385.

3. ***Merck Index***, 12th edn., S. Budavari, ed., Merck & Co., Whitehouse Station, NJ, 1996, p. 974.

4. C.R. Noller, ***Chemistry of Organic Compounds***, W.B. Saunders, Philadelphia, 1965, p. 887.

5. A. McKenzie, H.J. Plenderleith, and N. Walker, *J. Chem. Soc.*, 2875 (1923).

6. P. Newman, ***Optical Resolution Procedures for Chemical Compounds***, volume 2, part 1, Optical Resolution Information Center, Manhattan College, 1981, pp. 66-67.

7. A. Apelblat and E. Manzurola, *J. Chem. Thermodynamics*, **19**, 317 (1987).

8. H.G. Brittain, unpublished results.

9. L. Kryger and S.E. Rasmussen, *Acta Chem. Scand.*, **26**, 2349 (1972).

10. H. Landolt, ***Optical Rotating Power of Organic Substances***, English translation by J.H. Long, Chemical Publishing Co., Easton PA, 1902, pp. 528-536.

11. P. Walden, *Ber. Chem. Ges.*, **30**, 2889 (1897).

12. T.M. Lowry, ***Optical Rotatory Power***, Longmans, Green, and Co., London, 1935, pp. 293-298.

13. C. Djerassi, ***Optical Rotatory Dispersion***, McGraw-Hill, New York, 1960, p. 204.

14. H.G. Brittain, *Inorg. Chem.*, **19**, 2136 (1980); *Inorg. Chem.*, **20**, 959 (1981).

15. R.A. Copeland and H.G. Brittain, *Polyhedron*, **1**, 693 (1982); *J. Luminescence*, **27**, 307 (1982).

16. M. Ransom and H.G. Brittain, *Inorg. Chim. Acta*, **65**, L147 (1982)

17. J.F.J. Van Loock, W. Van Havere, and A.T.H. Leenstra, *Bull. Soc. Chim. Belg.*, **90**, 161 (1981).

18. C.J. Pouchert, ***Aldrich Library of Infrared Spectra***, 2nd edn., Aldrich Chemical Co., Milwaukee, WI, 1975, p. 275.

19. D.E. Bugay and W.P. Findlay, ***Pharmaceutical Excipients: Characterization by IR, Raman, and NMR Spectroscopy***, Marcel Dekker, New York, 1999, pp. 346-347.

20. S.-I. Sasaki, ***Handbook of Proton-NMR Spectra and Data***, volume 1, Academic Press, Tokyo, 1985, p.158

21. ***United States Pharmacopoeia 24***, the United States Pharmacopoeial Convention, Rockville, MD, 2000, p. 2475.

22. ***Official Methods of Analysis***, AOAC International, Arlington, VA, 1995, chapter 37, pp. 11-13.

NIMESULIDE

Amarjit Singh[1], Paramjeet Singh[1], Vijay K. Kapoor[2]

(1) Research & Development Centre
Panacea Biotec Ltd.
P.O. Lalru 140 501
India

(2) University Institute of Pharmaceutical Sciences
Panjab University
Chandigarh 160 014
India

1075-6280/01 $35.00

Contents

1. Description

1.1 Nomenclature

1.1.1 Systematic Chemical Name

4′-Nitro-2′-phenoxymethanesulphonanilide [1]

N-(4-Nitro-2-phenoxyphenyl)methanesulfonamide [2]

4-nitro-2-phenoxymethanesulfonanilide [2]

(Methylsulfonyl)(4-nitro-2-phenoxyphenyl)amine [3]

1.1.2 Nonproprietary Names [1, 2]

Nimesulide

1.1.3 Proprietary Names [1, 2, 4]

Algolider; Antifloxil; Aulin; Eudolene; Fansidol; Flogovital; Flolid; Guaxan; Laidor; Ledoren, Ledoven; Mesid; Mesulid; MF 110; Nide; Nidol; Nimedex; Nimesulene; Nims; Nimulid; Nisal; Nisulid; Remov; Resulin; Sulide; Teonim

1.2 Formulae

1.2.1 Empirical Formula, Molecular Weight, CAS Number

$C_{13}H_{12}N_2O_5S$ [MW = 308.31]

CAS number = 51803-78-2

1.2.2 Structural Formula

H SO$_2$CH$_3$
N
O
NO$_2$

1.3 Elemental Analysis

The calculated elemental composition is as follows:

carbon:	50.64%
hydrogen:	3.92%
oxygen:	25.95%
nitrogen:	3.92%
sulfur:	10.40%

1.4 Appearance

Nimesulide is a light yellow crystalline powder, which is practically odorless.

1.5 Uses and Applications

Nimesulide has analgesic, anti-inflammatory, and antipyretic properties, acting as an inhibitor of prostaglandin synthetase and platelet aggregation. It is given in doses of up to 200 mg twice daily by mouth for inflammatory conditions, fever, and pain.

2. Methods of Preparation

Nimesulide has been synthesized by Moore and Harrington [5] in 1974. The procedure involved the dissolution of 2-phenoxy-methane sulfonanilide in glacial acetic acid (by warming), followed by the dropwise addition of an equimolar amount of 70% nitric acid with continuous stirring. After heating the mixture on a steam bath for four hours, the final product is obtained by pouring the mixture in water. At this point, Nimesulide is precipitated.

$NHSO_2CH_3$ O $\xrightarrow[\text{Acetic Acid}]{70\%\ HNO_3}$ $NHSO_2CH_3$ O NO_2

An alternate route of synthesis (Scheme 2) has been developed [6] which starts with *o*-chloronitrobenzene and phenol. Nimesulide is reported [7] to be prepared in four steps in 61% overall yield from *o*-chloronitrobenzene and phenol.

OH

KOH

OK

NO_2

Cl

Cu-Catalyst

NH_2

O

Fe/HCl

NO_2

O

CH_3SO_2Cl $(C_2H_5)_3N$

$NHSO_2CH_3$

O

70% HNO3

CH_3COOH

$NHSO_2CH_3$

O

NO_2

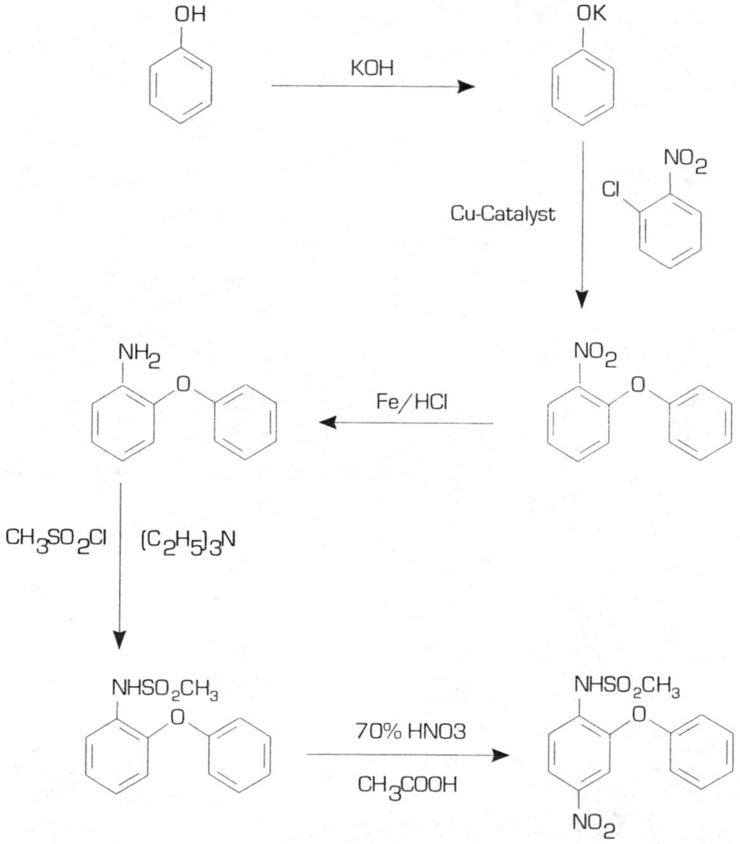

A series of compounds structurally related to Nimesulide have been prepared in which *p*-nitrophenyl moiety has been replaced by pyridine and pyridine *N*-oxide [8]. In another modification, the p-nitro group of Nimesulide was substituted by a cyano and 1*H*-tetrazol-5-yl groups. Analogs were also prepared where the methanesulfonamido group was replaced by an acetamido group. However all such modifications were found to be detrimental to the activity.

Nimesulide analogs have been prepared that exhibit inhibition of bladder cancer in animal models [9], and as prophylactics and therapeutics for ischemia-reperfusion injury [10]. A number of Nimesulide analogs that act as anti-inflammatory agents have also been reported [11-17].

3. Physical Properties

3.1 Ionization Constants

Nimesulide is characterized by a single ionization constant associated with dissociation of the -NH proton of the sulfonanilide group. Various pKa values have been reported in the literature: 5.9 [38], 6.46 [39], 6.50 [40], and 6.56 [41]. These values clearly indicate the acidic nature of the drug.

A method for automated pKa determination by capillary zone electrophoresis has been described [42].

3.2 Solubility Characteristics

Nimesulide is soluble in moderately polar solvents such as dichloromethane and acetone. The solubility is diminished in solvents of high polarity such as methanol. The solubility of Nimesulide in water is reported to be 0.01 mg/mL [22], which becomes enhanced by an increase in the pH of the aqueous solution. This is essentially due to deprotonation and ionization of sulfonanilide group. Table 1 provides the solubility characteristics of Nimesulide in buffers over the pH range of 4 to 11, as well as in a variety of solvents.

To overcome the problem of very poor water solubility, various inclusion complexes with β-cyclodextrin [23-28] and hydroxypropyl β-cyclodextrin [28] have been prepared. A Nimesulide-β-cyclodextrin complex has been reported to have an aqueous solubility of about 16 mg/mL [22]. To assess the efficacy and tolerability of single doses of Nimesulide-β-cyclodextrin, a comparative study of the complex and Nimesulide has been carried out in patients with dental pain following surgical procedures [29]. *In-vitro* and *in-vivo* studies on the sodium Nimesulide-β-cyclodextrin inclusion complex have been carried out [30].

Table 1

Solubility Characteristics of Nimesulide

Solvent System	**Solubility (mg/mL)**
Buffer pH 4.0	0.068
Buffer pH 5.0	0.0078
Buffer pH 6.0	0.0091
Buffer pH 7.0	0.0021
Buffer pH 8.0	0.152
Buffer pH 9.0	0.666
Buffer pH 10.0	1.03
Buffer pH 10.5	1.41
Buffer pH 11.0	2.03
Methanol	9.43
Dichloromethane	163.0
Acetone	162.4
Acetonitrile	92.7
Ethanol (95%)	3.6
Ethyl acetate	89.3

Since the Nimesulide-L-lysine salt exhibited increased aqueous solubility (5-7.5 mg/mL), Nimesulide-L-lysine-β or γ-cyclodextrin complexes were prepared by spray-drying [22, 31]. These complexes yielded remarkable aqueous solubilities. The incorporation of Nimesulide in a Nimesulide-L-lysine-β-cyclodextrin complex increased the water solubility by a factor of 10 at pH 1.5, and 160 at pH 6.8. A 3600-fold increase in solubility was reported in purified water (36.40 mg/mL for the complex, as compared to 0.01 mg/mL for Nimesulide alone) [22].

Solid dispersions of Nimesulide in polyvinyl pyrrolidone, polyethylene glycol, dextrin, and pregelatinized starch have been prepared and evaluated for dissolution rates [32, 33].

A water-soluble adduct of Nimesulide with an amino-sugar, *N*-methyl-glucamine, has been patented for injectable use [34]. Nimesulide micronized salts with metals such as sodium, potassium, calcium, magnesium, and zinc having improved bioavailability and pharmacokinetics have been prepared [35]. A Nimesulide choline salt has also been prepared [36].

The solubility of Nimesulide in supercritical carbon dioxide, measured using a dynamic saturation technique of pressures between 100 bar and 220 bar and at two temperatures, 312.5 K and 331.5 K, has been reported [37].

3.3 Partition Coefficient

The octanol/water partition coefficient of Nimesulide is 238, corresponding to a log P value of 2.376 [38]. A value of this magnitude clearly demonstrates the lipophilic character of the drug.

3.4 Particle Morphology

It is reported [2] that light-tan crystals are obtained from ethanol on crystallization. A commercial sample crystallized from ethanol yielded needle-shaped crystals [18]. The two samples were evaluated using electron microscopy, with the data being obtained on a JEOL JSM-6100 system (Figures 1a and 1b) [18]. As evident in the photomicrograph, the majority of the crystals exhibited a rod-like morphology.

Figure 1. Scanning electron photomicrographs obtained at a magnification of 2000X for (a) a commercial sample of Nimesulide, and (b) a recrystallization of the commercial sample.

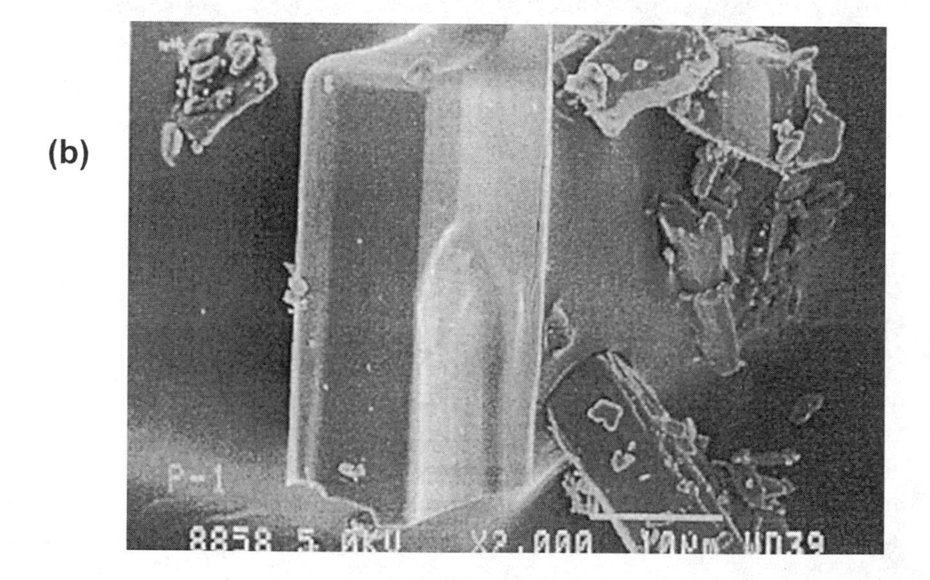

Most of the crystals of the commercial sample were approximately 5 to 20 μm in length and 1 to 5 μm in width, whereas recrystallized commercial samples were more than 100 μm in length and 10-12 μm in width. One concludes from consideration of Figure 1 that the commercial sample is micronized before being used to produce dosage forms.

3.5 Crystallographic Properties

3.5.1 Single Crystal Structure

A full single crystal structural determination of Nimesulide has been reported [19]. The compound crystallizes in the monoclinic space group $C2/_C$, with $a = 33.657(3)$ Å, $b = 5.1305(3)$ Å, $c = 16.0816(10)$ Å, and $\beta = 92.368\ (8)°$. In addition, Z = 8, $d_c = 1.476$; $K(F^2) = 0.0401$, $R_w\ (F^2) = 0.1146$ for 1908 reflections. The molecular conformation is stabilized by an intramolecular N-H···O hydrogen bond. The angle between the two phenyl rings is 74.7°. The cohesion of the crystal is the result of N-H···O intermolecular hydrogen bonds and van der Waals interactions. Figure 2 gives the structure derived from the data.

There is no definite information that conclusively demonstrates the existence of crystal polymorphs. There is a report, however, indicating a faster dissolution rate for crystals prepared by solvent change (1:1 ethanol / water) in the presence of 1% Tween 80 compared to those crystals in presence of polyethylene glycol 4000 and povidone K30 [20]. Adsorption of the surfactant or the water-soluble polymer (PEG or PVP) on the crystal surface could possibly account for any faster crystal dissolution.

3.5.2 X-Ray Powder Diffraction Pattern

The X-ray powder diffraction pattern of Nimesulide was obtained using a Philips PW1729 X-ray diffractometer system [18]. The radiation source was a copper (λ= 1.54820Å) high intensity X-ray tube operated at 35 kV and 20 mA.

Nimesulide was found to exhibit a strong and characteristic x-ray powder diffraction pattern, showing the crystalline nature of the powder. The powder pattern is found in Figure 3, and Table 2 provides a summary of scattering angles, d-spacings, and relative intensities.

Figure 2. Molecular conformation of Nimesulide.

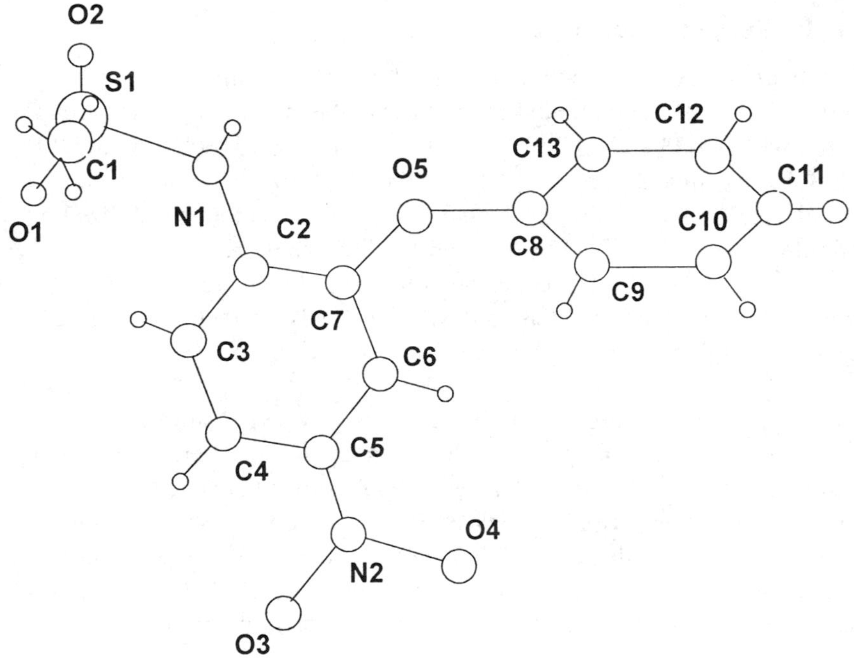

Figure 3. X-ray powder diffraction pattern of Nimesulide.

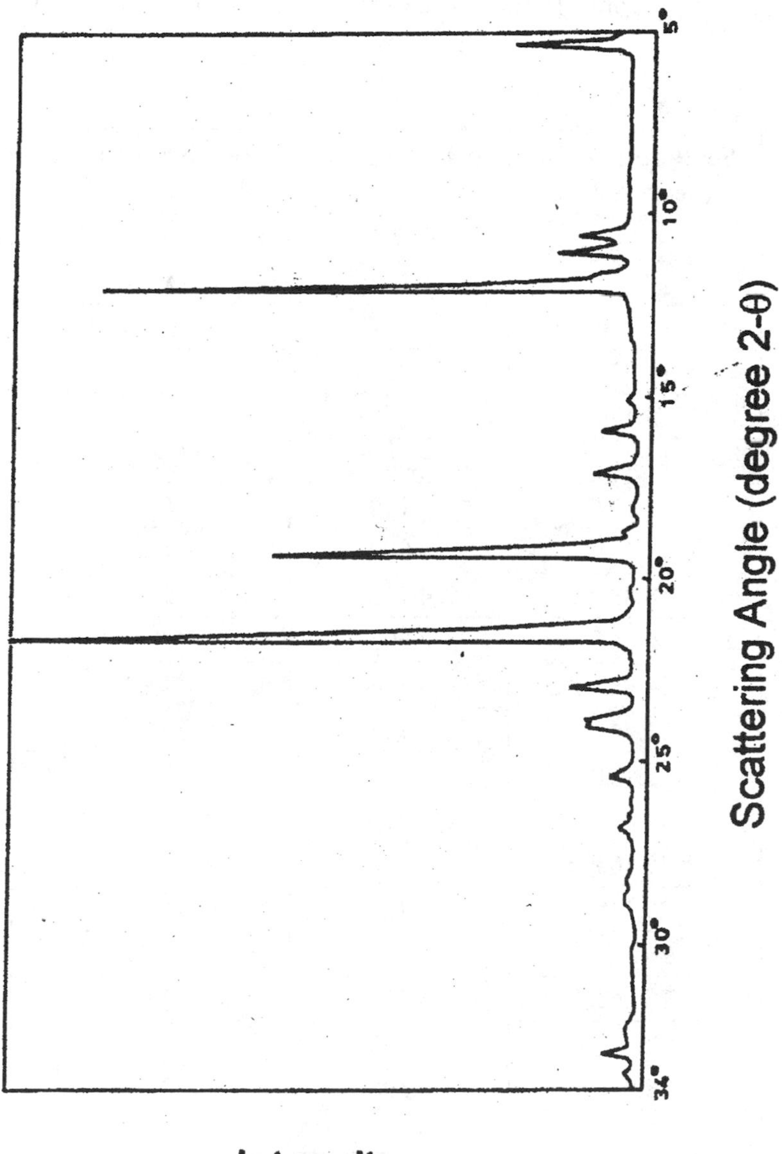

Table 2

Crystallographic Parameters Derived from the X-Ray Powder Pattern of Nimesulide

Scattering Angle (degrees 2-θ)	**d-spacing (Å)**	**Relative Intensity, I/I_0 (%)**
6.291	16.771	16.04
10.633	8.354	8.44
11.110	7.997	11.07
12.032	7.386	67.62
15.139	5.876	2.46
15.989	5.565	5.84
17.157	5.189	6.60
18.145	4.909	1.96
18.829	4.732	2.81
19.349	4.606	37.19
20.383	4.374	1.98
21.646	4.122	100.00
23.121	3.862	8.94
24.027	3.719	7.05
25.500	3.507	4.09
26.145	3.422	1.64
26.579	3.367	1.83
26.862	3.332	3.08
28.302	3.166	1.92
28.916	3.198	2.06
29.009	3.090	1.95
30.274	2.964	1.21

Although the powder pattern contains a number of scattering lines, the pattern is dominated by the intense scattering peaks located at 12.03, 19.34, and 21.6 degrees 2-θ.

3.6 Hygroscopicity

When exposed to relative humidities ranging from 60% to 70% (at an equilibrium temperature ranging from 25°C to 40°C), Nimesulide did not exhibit any measurable moisture pickup in over seven days. The compound is therefore determined to be non-hygroscopic.

3.7 Thermal Methods of analysis

3.7.1 Melting Behavior

The melting range of Nimesulide is 148 – 150°C, and the substance appears to melt without decomposition. Other reported values in the literature for melting range are 143 – 144.5°C [2] and 149°C [21].

3.7.2 Differential Scanning Calorimetry

The differential scanning calorimetry thermogram of Nimesulide was obtained using a Mettler-Toledo model DSC821, calibrated using 99.999% indium [18]. The thermogram shown in Figure 4 was obtained using a heating rate of 5°C/min, under an atmosphere of nitrogen. A single sharp endotherm was observed, having an onset of 148.71°C, a maximum at 149°C, and an endset at 151.67°C. The endotherm is assigned to the melting of the compound, and is characterized by an enthalpy of fusion equal to 650 Joule/g (200.4 kJ/mol). The quality of the thermogram indicates that DSC could be used as one of the techniques to determine the purity of Nimesulide.

3.7.3 Thermogravimetric Analysis

As an anhydrous material, Nimesulide exhibits no loss in weight until heated beyond the temperature of its thermal decomposition (approximately 300°C). The TG thermogram is shown in Figure 5 [18].

Figure 4. Differential scanning calorimetry thermogram of Nimesulide.

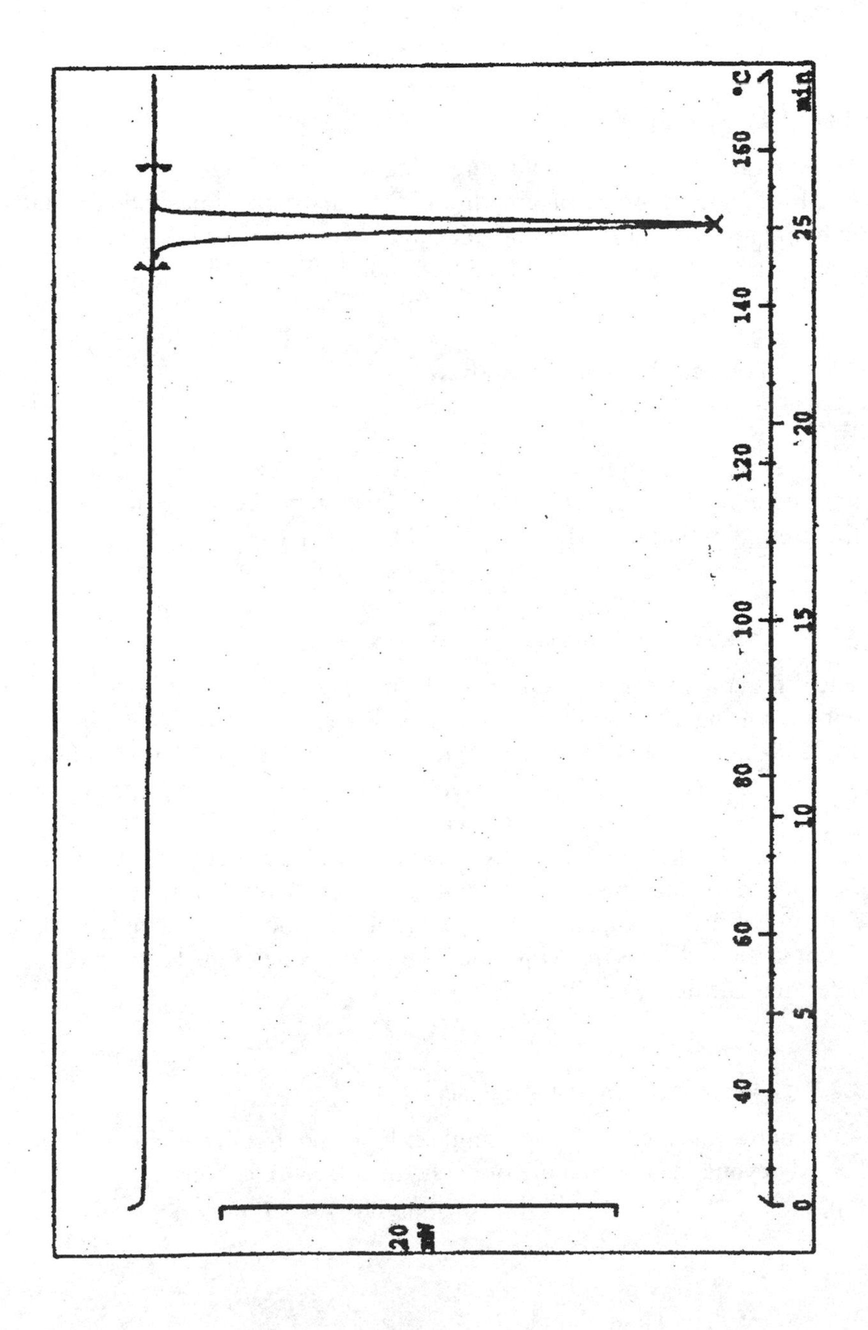

Figure 4. Thermogravimetric analysis thermogram of Nimesulide.

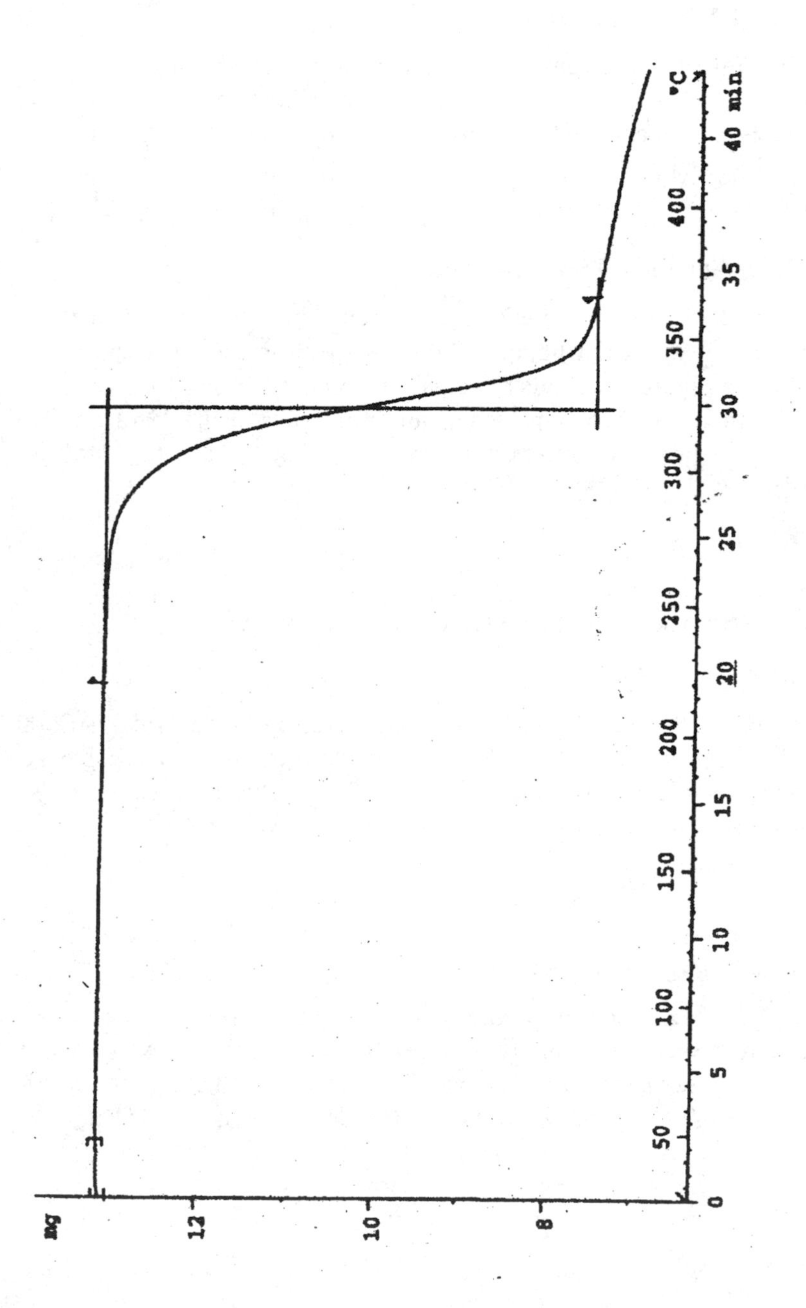

3.8 Spectroscopy

3.8.1 UV/VIS Spectroscopy

The ultraviolet spectrum of Nimesulide was recorded using a UV-Beckman DU 640i spectrophotometer, at a solute concentration of 0.04 mg/mL in methanol. The spectrum shown in Figure 6 exhibits maxima at 213, 296, and 323 nm [18].

3.8.2 Vibrational Spectroscopy

The infrared spectrum of Nimesulide obtained in a KBr pellet, and was recorded on a-Nicolet-Impact-400) FTIR spectrometer. The spectrum shown in Figure 7 exhibits principal stretching modes at 1593 cm^{-1} (aromatic rings), 1153 cm^{-1} (SO_2 antisymmetric stretch), 1342 cm^{-1} and 1520 cm^{-1} (aryl nitro group stretching), 1247 cm^{-1} (diaryl C-O stretching), and 3288 cm^{-1} (NH stretch) [18].

3.8.3 Nuclear Magnetic Resonance Spectrometry

3.8.3.1 ^{1}H-NMR Spectrum

The ^{1}H-NMR spectra of Nimesulide were recorded on a Bruker AC300F NMR spectrometer at 300 MHz, using deutero-chloroform as solvent and tetramethylsilane as the internal standard [18]. The spectra are shown in Figures 8 and 9, and Table 3 provides a summary of the proton assignments.

3.8.3.2 ^{13}C-NMR Spectrum

The ^{13}C-NMR spectrum (Figure 10) of Nimesulide was obtained in deutero-chloroform at ambient temperature using tetramethylsilane as the an internal standard [18]. The one-dimensional and the DEPT 135 (Figure 11), DEPT 90 (Figure 12), COSY 45 (Figure 13), and INVBTP (HETEROCOSY) (Figure 14) spectra were used to develop the ^{13}C chemical shift assignments that are summarized in Table 4.

Figure 6. Ultraviolet spectrum of Nimesulide in methanol.

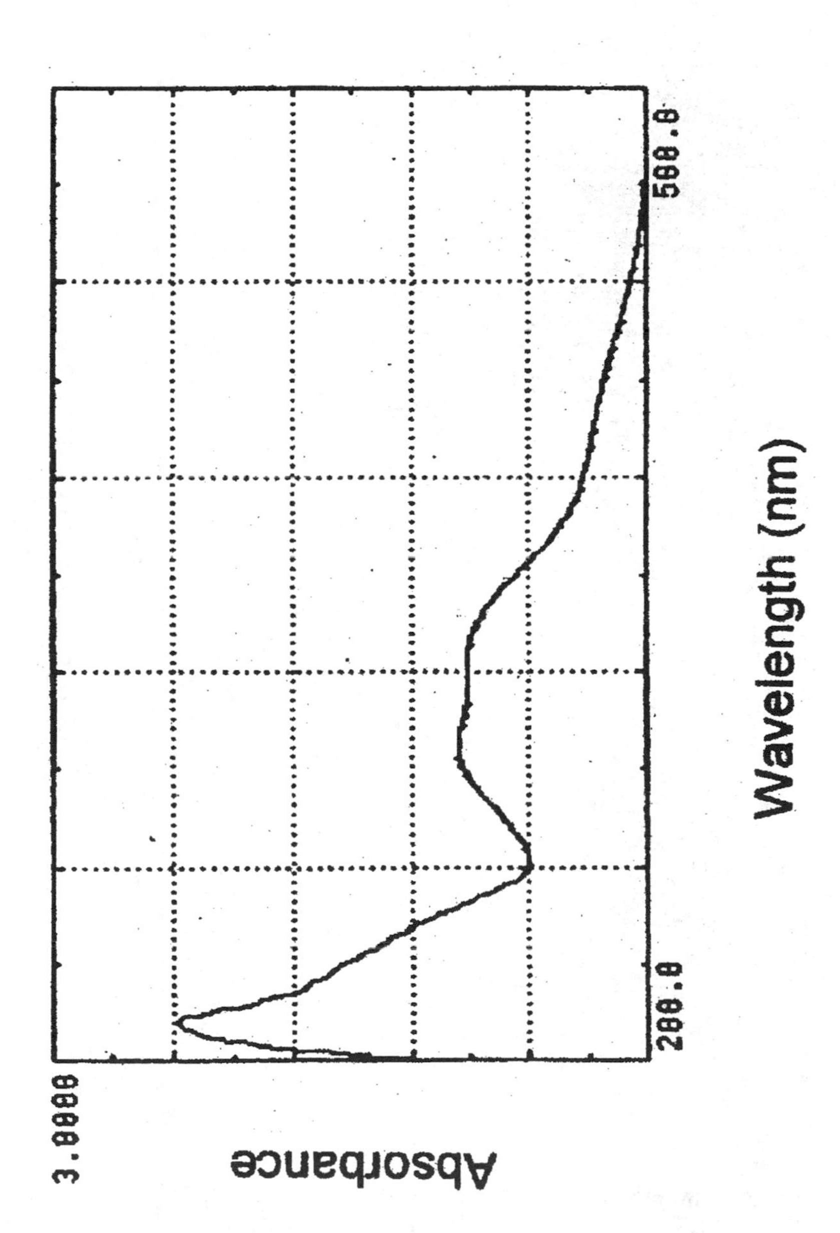

Figure 7. Infrared absorption spectrum of Nimesulide in a KBr pellet.

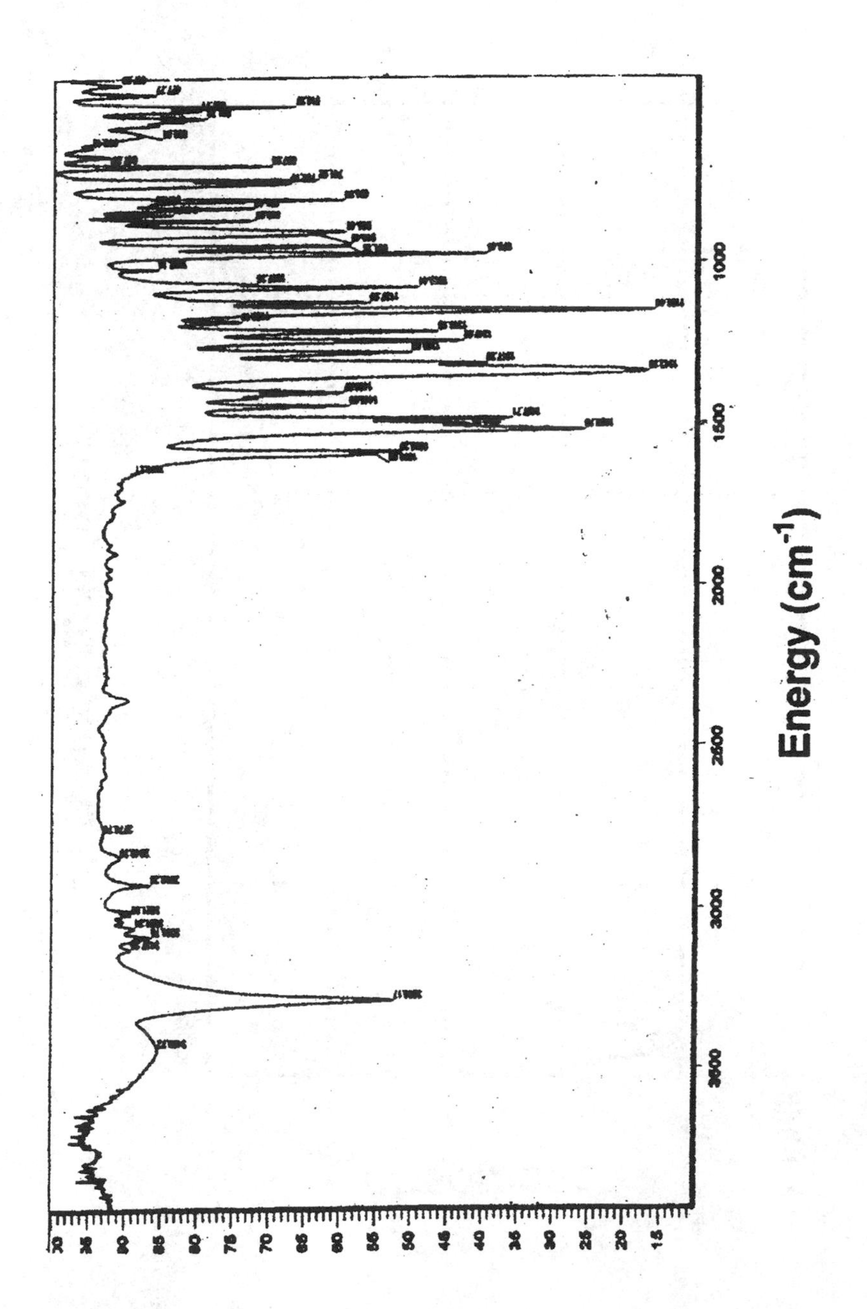

Figure 8. ^{1}H-NMR spectrum of Nimesulide in $CDCl_3$.

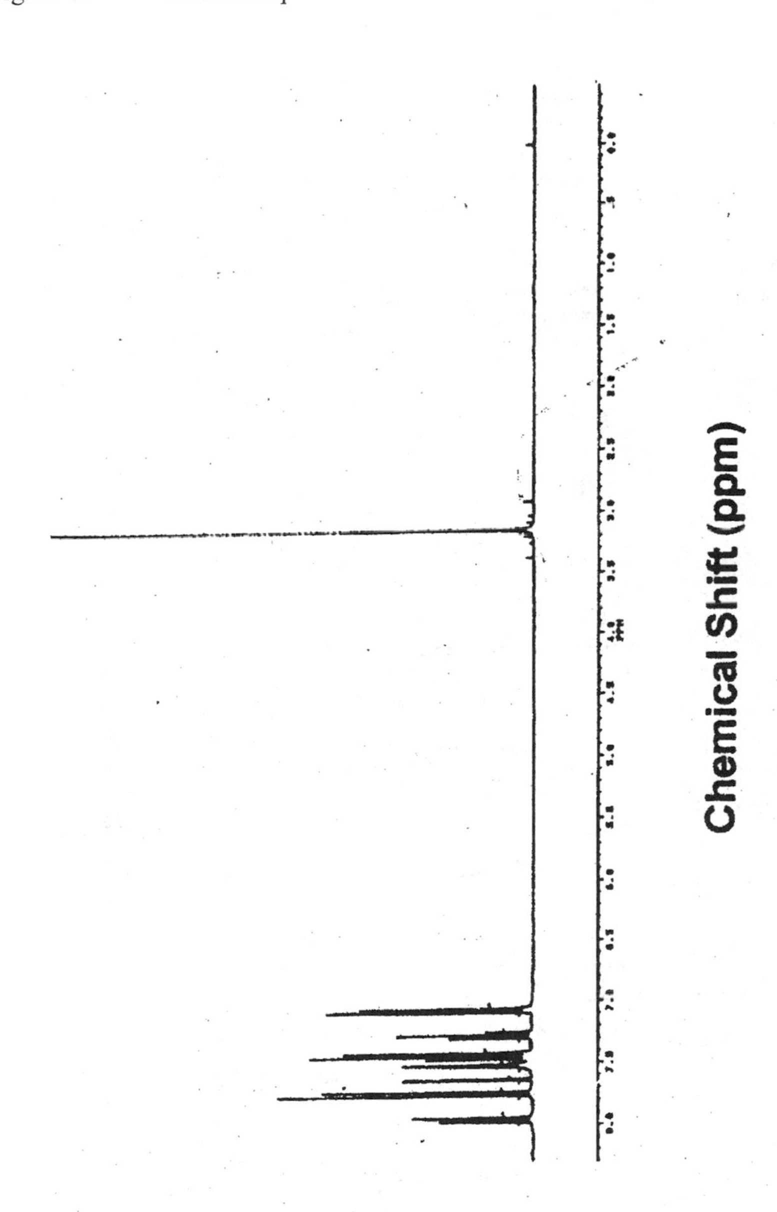

Figure 9. Expanded ^{1}H-NMR spectrum of Nimesulide in $CDCl_3$.

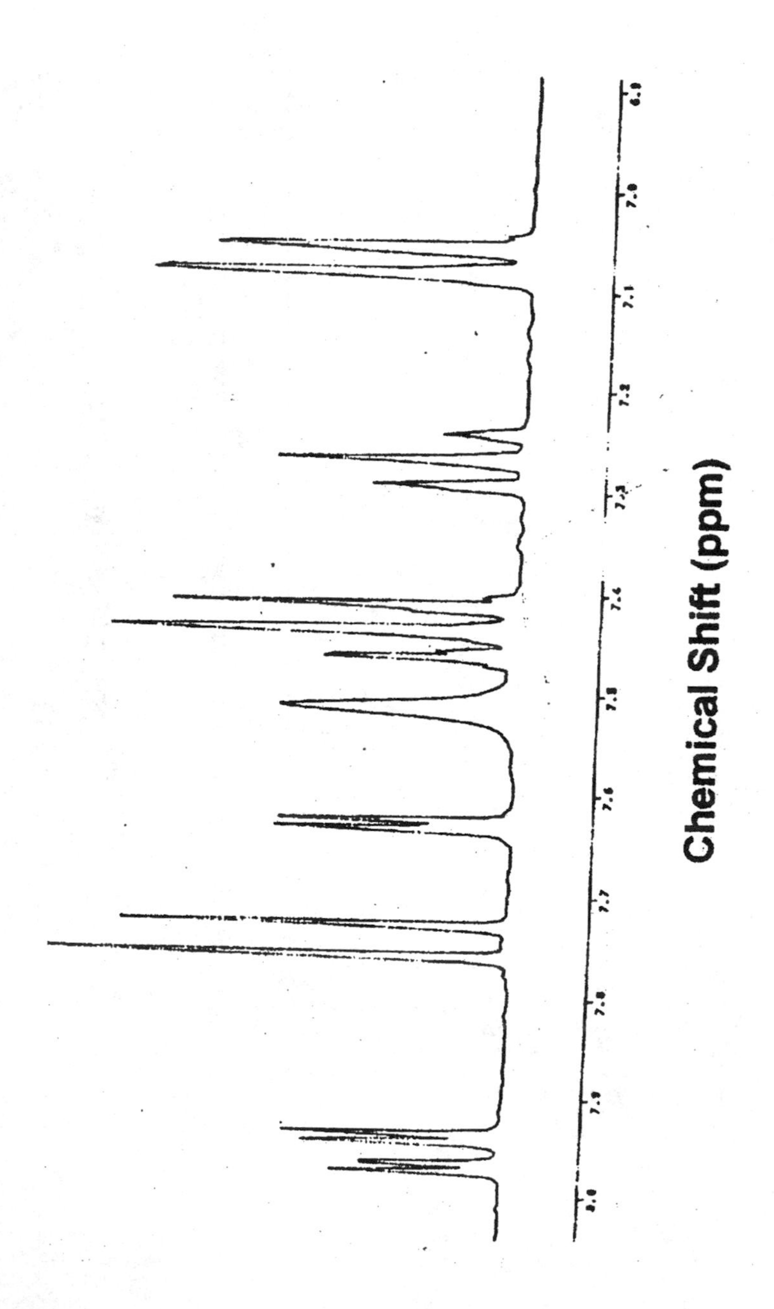

Table 3

Assignments for the Resonance Bands Observed in the ^{1}H-NMR Spectrum of Nimesulide

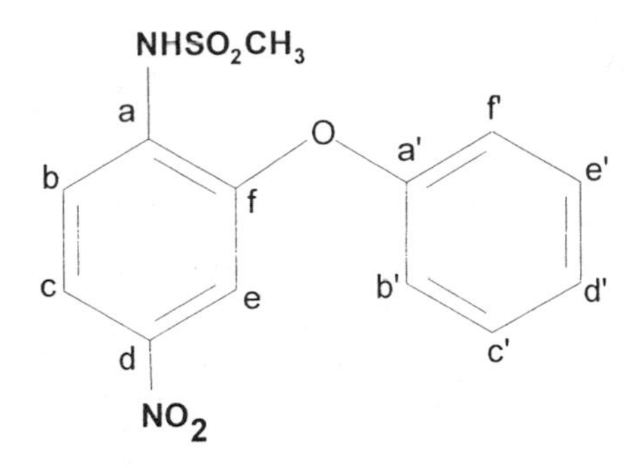

Chemical Shift (ppm)	Number of Protons	Multiplicity	Assignment
3.14	3	Singlet	$-SO_2CH_3$
7.05-7.08	2	Doublet	b';f'; Ar-H
7.24-7.29	1	Triplet	d'; Ar-H
7.41-7.46	2	Triplet	e'; c'; Ar-H
7.51	1	Singlet	$-NH-SO_2-$
7.61	1	Doublet	e; Ar-H
7.73-7.75	1	Doublet	b; Ar-H
7.93-7.97	1	Double-doublet	c; Ar-H

Figure 10. ^{13}C-NMR spectrum of Nimesulide in $CDCl_3$.

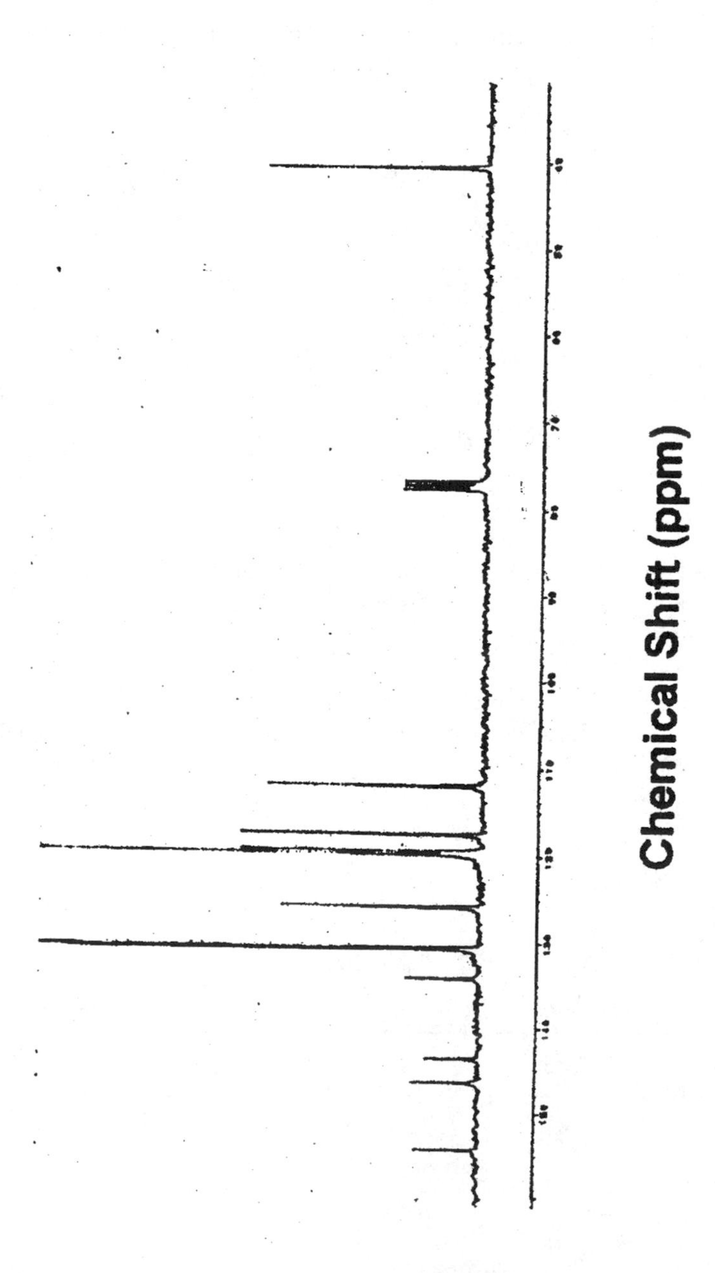

Figure 11. DEPT 135 ^{13}C-NMR spectrum of Nimesulide.

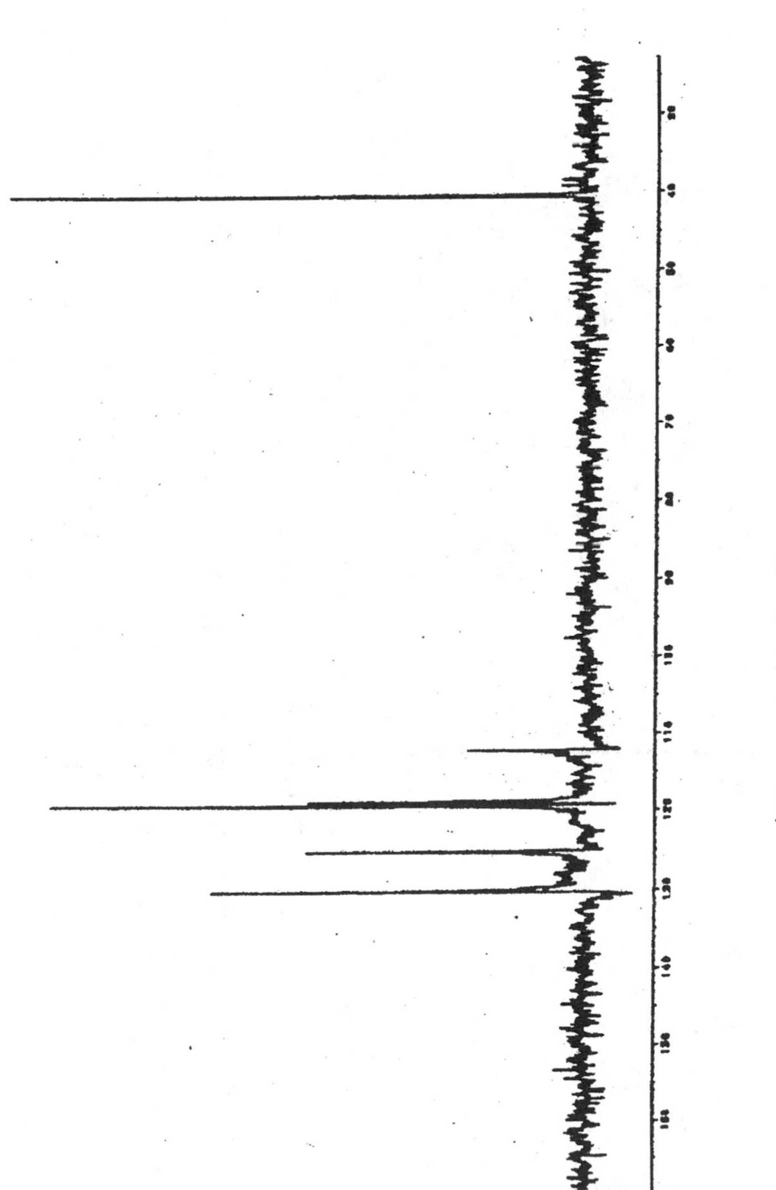

Figure 12. DEPT 90 ^{13}C-NMR spectrum of Nimesulide.

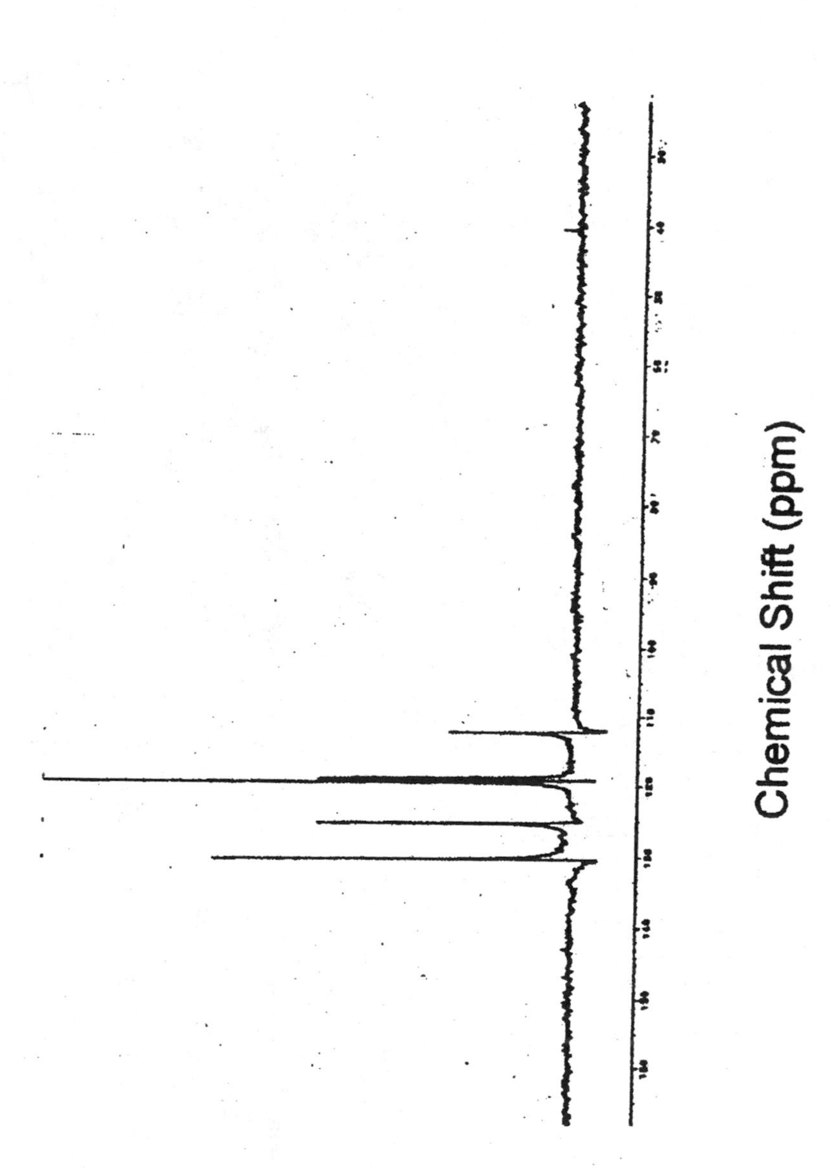

Figure 13. HOMOCOSY NMR spectrum of Nimesulide.

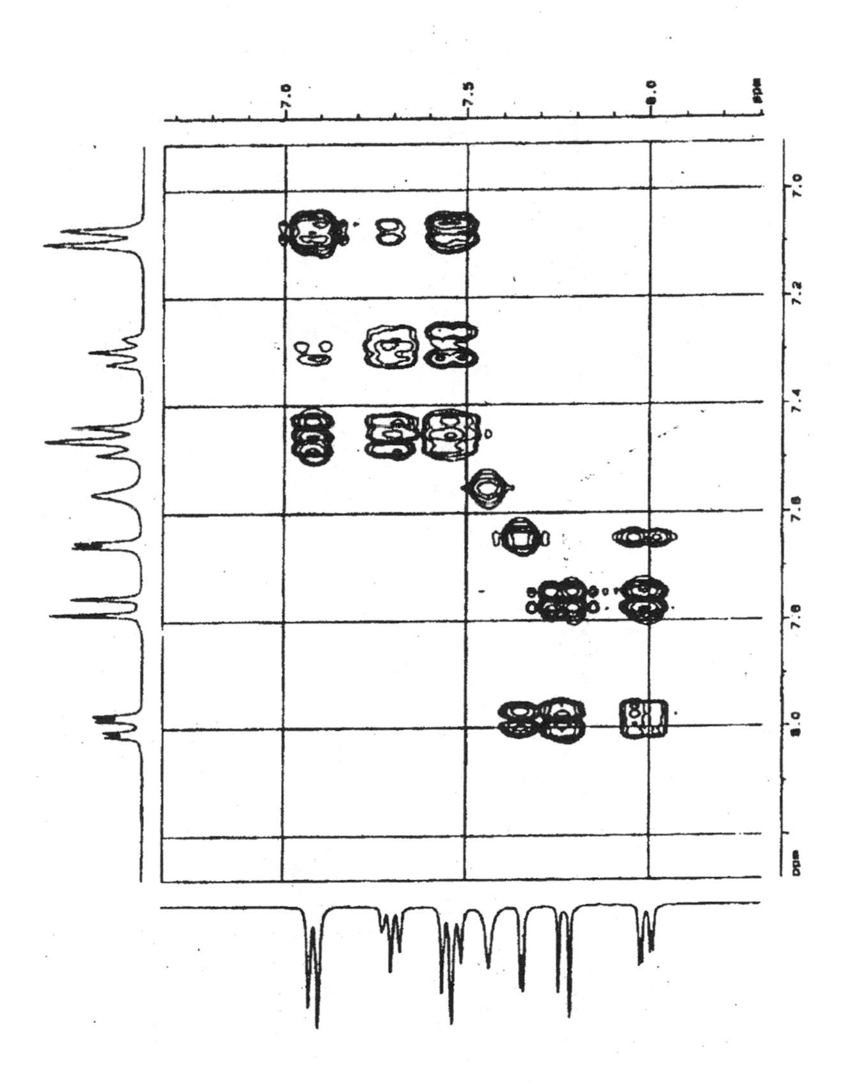

Figure 14. HETEROCOSY NMR spectrum of Nimesulide.

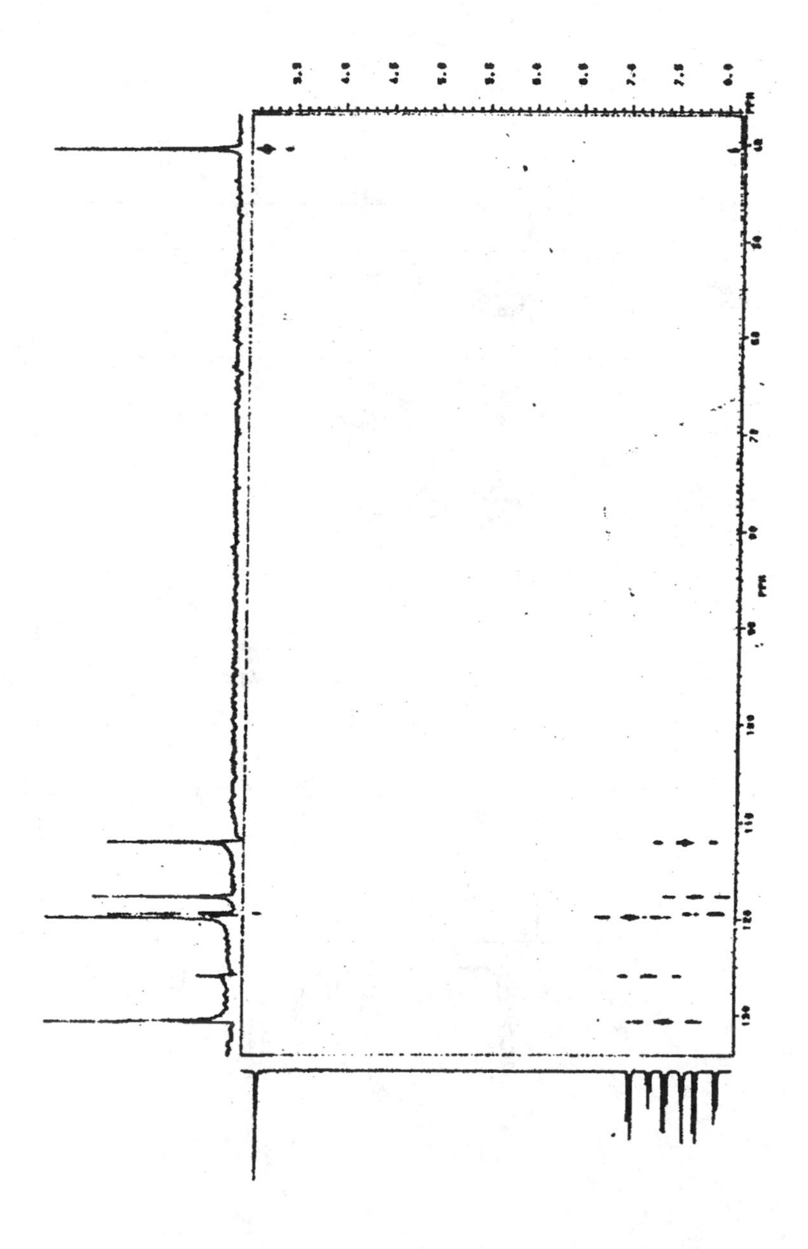

Table 4

Assignments for the Resonance Bands Observed in the ^{13}C-NMR Spectrum of Nimesulide

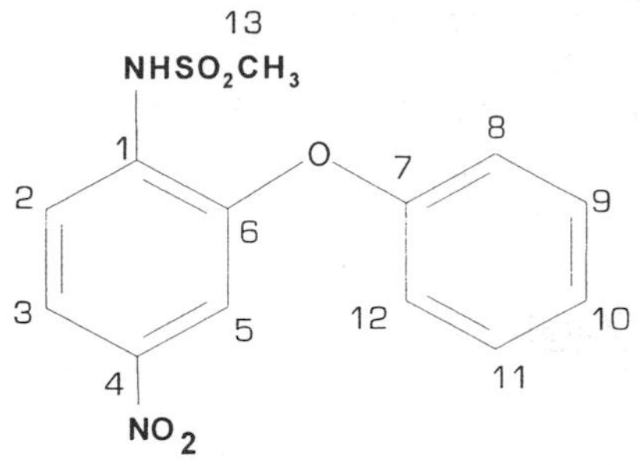

13
$NHSO_2CH_3$
1 2 3 4 5 6 O 7 8 9 10 11 12
NO_2

Chemical Shift (ppm)	Carbon Number
134.08	1
117.50	2
119.33	3
154.20	4
111.89	5
146.38	6
143.59	7
119.66	8,12
130.58	9,11
125.79	10
40.46	13

3.9 Mass Spectrometry

The mass spectrum of Nimesulide was recorded using a VG 70-250S mass spectrometer, and is shown in Figure 15 [18]. The molecular ion peak (M^+) was found at m/z = 308 (80%). The other characteristic peaks appeared at:

m/z = 229 (100%), due to M^+ - SO_2CH_3
m/z = 183 (37.4%), due to M^+ - (SO_2 CH_3 - NO_2)
m/z = at 154 (99.2%)
m/z = 77 (47.2%), due to $C_6H_5^+$

4. Methods of Analysis

4.1 Identification

The identification of Nimesulide can be made on the basis of its characteristic ultraviolet absorbance, or on the basis of its infrared absorption spectrum. A method for identification based on thin layer chromatography will be discussed in a later section.

4.2 Titrimetric Analysis

A titrimetric method has been reported for the assay of Nimesulide [3]. 0.240 g of sample is dissolved in 30 mL of acetone, to which 20 mL of water is added. The solution is titrated with 0.1M sodium hydroxide, and the end point is determined potentiometrically. Each milliliter of 0.1 M NaOH is equivalent to 30.83 mg of Nimesulide.

Potentiometric titrations have also been employed for the pKa determination of Nimesulide in methanol-water mixtures [39].

4.3 Electrochemical Analysis

Several electrochemical methods for the analysis of Nimesulide have been reported [43-45]. Nimesulide in hydro-alcoholic solution presents a cathodic response over a wide range of pH values (2-12), both by differential pulse and polarographic techniques. The results show only one well-defined peak or wave over all of the pH range studied. This wave corresponds to reduction of the nitro group at position 4.

Figure 15. Mass spectrum of Nimesulide.

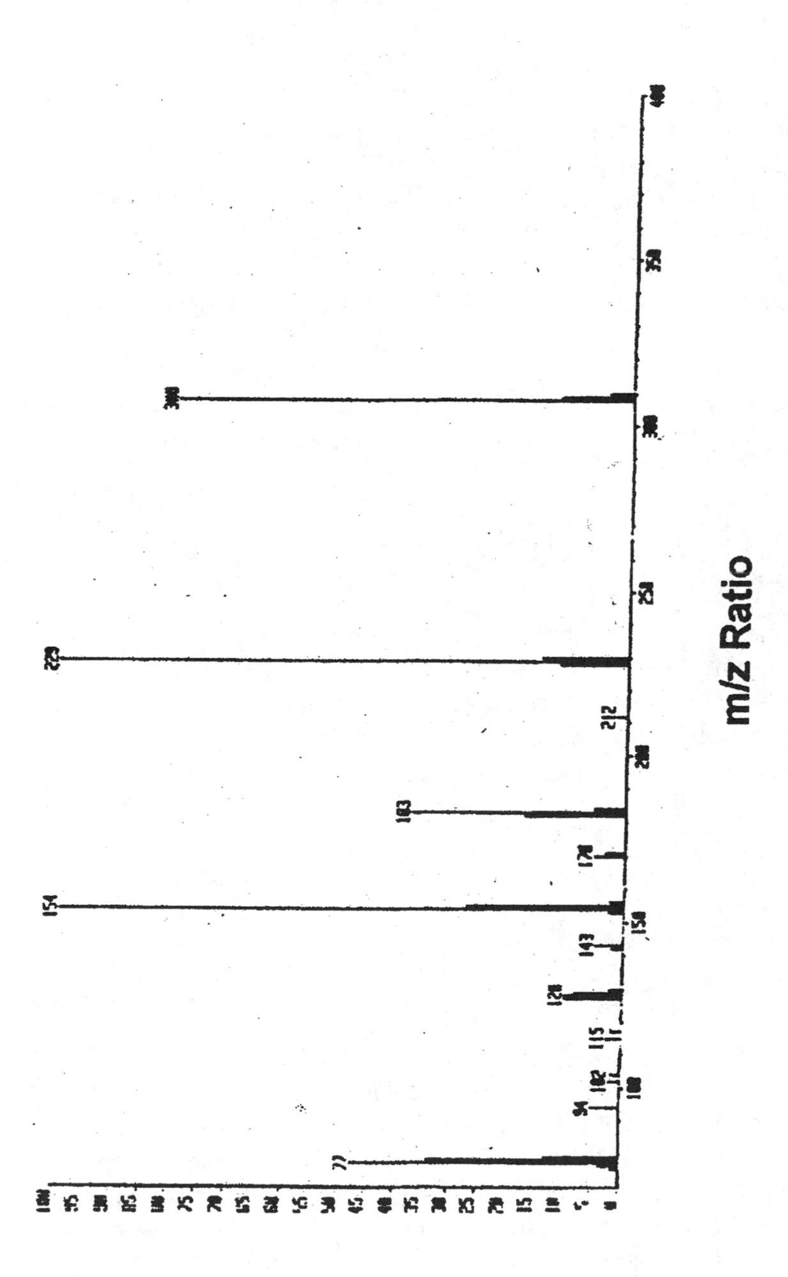

The voltammetric oxidation shows one well-resolved signal within the pH range studied. This anodic signal could be attributed to the menthanesulfonamide group oxidation. For analytical purposes, a very well resolved diffusion controlled differential pulse polarographic peak obtained at pH 7.0 was selected. This peak was used to develop a new method for the determination of Nimesulide in pharmaceutical dosage forms. The recovery was 104.8%, with a RSD of 1.3%. The method was reported to be sufficiently accurate and precise so as to be applied in the individual tablet assay of commercial samples [43].

A simple D.C. polarographic method has been developed for the analysis of nitro-containing drugs (such as Nimesulide) in tablets. The optimum pH range for obtaining well resolved waves suitable for quantitative determination of the drug was found to be between 4.0 and 6.0. Both the standard addition and calibration methods were employed [45].

4.4 Spectroscopic Analysis

4.4.1 Fluorimetry

Nimesulide was determined by a fluorimetric method based on quenching the natural fluorescence of *N*-(1-naphthyl)ethylenediamine by reaction with the diazotized reduction product of Nimesulide [46]. In a methanolic solution, the drug was reduced by zinc dust in 1M hydrochloric acid and diazotized with sodium nitrite and hydrochloric acid. After the addition of sodium sulfamate and *N*-(1-naphthyl)ethylenediamine, the fluorescence was measured at 309/427 nm (excitation/emission). Beer's law was obeyed over the concentration range of 0.55 - 2.75 μg/mL, and the analyte recovery was 99.5 - 100.4%. Common tablet excipients did not interfere in the assay [46].

4.4.2 Colorimetry and Spectrophotometry

Several spectrophotometric methods have been reported for the determination of Nimesulide in pharmaceutical dosage forms [47-56]. Fallavana *et al.* [47] have reported an ultraviolet spectrophotometric method, based on the absorbance at 295 nm, for the determination of Nimesulide in tablets. The concentration of the working curve was 9.6 - 30.4 mg/mL, and Pearson's coefficient was 0.99978.

A spectrophotometric method has been described which is based on reduction of the nitro group of Nimesulide by zinc dust and hydrochloric acid, followed by diazotization and coupling with *N*-(1-naphthyl)ethtylene-diamine dihydrochloride, to form a stable purple chromophore absorbing at 557 nm [48, 50]. A simple spectrophotometric method based on formation of a blue species (λ_{max} = 600 nm) with Folin-Ciocalteu reagent has been described [51]. Methods based on the formation of a colored product from the reaction of Nimesulide with paradimethylamino-cinnamaldehyde (exhibiting maximum absorption at 525 nm), and on complex formation with 1,10-phenanthroline and Fe(III) (with maximum absorption at 490 nm) have been reported [52].

A rapid, accurate and simple method for determination of bulk Nimesulide, and in its dosage forms, based on the formation of a colored product with vanillin, has been reported [53]. Beer's law was obeyed over the range of 33.3 - 166.6 μg/mL, with an RSD of 0.38%. A bluish green chromophore was formed when reduced Nimesulide was reacted with 3-methyl-2-benzothiazolinonehydrazone hydrochloride in the presence of ferric chloride(absorption maximum of 600 nm). This latter method was employed for the analysis of tablets and suspensions containing Nimesulide [54]. The recovery was 99 - 101%.

In another method [55], formation of a yellow solution having maximum absorbance at 436 nm, was obtained on dissolving Nimesulide in 0.1 N sodium hydroxide. This was the basis for determination of the drug in pharmaceutical dosage formulations. Manna *et al.* [56] have reported formation of a greenish yellow chromophore in sodium hydroxide solution (pH 12) with Nimesulide, which exhibited a λ_{max} = 394 nm.

4.5 Bioassay

Bioassay techniques for non-steroidal anti-inflammatory drugs including Nimesulide have been reported [58-60]. When freshly drawn, heparinized human whole blood is incubated with 50 μM calcium ionophore A23187, and platelets are stimulated to produce thromboxane B_2 (Tx B_2) by activation of prostaglandin G/H synthase-1 (PGHS-1). Tx B_2 concentration, as measured by immunoassay, is maximal at 20 - 30 minutes and declines thereafter. Addition of non-steroidal anti-inflammatory drug prior to 30 minute stimulation with ionophore results in concentration dependent inhibition of Tx B_2 production [58]. *In-vivo* assessment of Nimesulide cyclooxygenase-2 selectivity has been reported by Shah *et al.* [59]. An *ex vivo* assay to determine the cyclooxygenase selectivity of non-steroidal anti-inflammatory drugs including Nimesulide has been described [60].

4.6 Chromatographic Methods of Analysis

4.6.1 Thin Layer Chromatography

A test sample of Nimesulide complies to the identification test if the retention factor of the sample is identical to that of an authentic Nimesulide reference standard.

A thin layer chromatography method has been developed for Nimesulide which is based on the use of silica gel 60 GF_{254} (Merck) as the stationary phase, and 9:1:1 toluene / ether / ethyl acetate as the mobile phase. The sample is prepared by weighing 10 mg of drug substance, and dissolving in 1 mL of methanol. 5 μL of this solution is applied to the plate and dried. The plate is allowed to run to a height of approximately 12 cm, and then allowed to dry. The spot is viewed using short wave ultraviolet (254 nm), and the retention factor (R_f) for Nimesulide is 0.427. R_f values for 2-phenoxy-phenylmethanesulfonanilide (a process impurity) and for 4-amino-2-phenoxymethanesulfonanilide (the reduced product of Nimesulide) are 0.558 and 0.175 respectively [18].

4.6.2 Gas Chromatography

A gas chromatographic method, using a 10% SE-30 column, has been reported for the determination of Nimesulide in its pharmaceutical preparations [61].

4.6.3 High Performance Liquid Chromatography

Several high performance liquid chromatography (HPLC) systems for the determination of Nimesulide have been reported [62-69]. The details are summarized in Table 5

Table 5

HPLC Methods for the Determination of Nimesulide

Column	Mobile phase	Flow rate (mL/min)	Detection	Ref.
C18 reversed-phase column (0.46 x 30 cm; 10 μm particle size)	0.05 M phosphate buffer (pH 5.0) – methanol (50 : 50)	1	UV at 230 nm	62
μ-Bondapack C18 reversed - phase column	0.05 M phosphate buffer (pH 7.0)-methanol (45 : 55)	1	UV at 230 nm	63
ODS column	Water (1g/L Na_2HPO_4) – methanol (30 :70)	1	UV at 254 nm	64
μ Bondapack/μ Porasil C18 column	Methanol-0.05 M phosphate buffer (pH 3)	1	Electro chemical	65
Hypersil ODS2 column	Methanol-water-acetic acid (60:40:1)	1	UV at 230 nm	66
Supelcosil LC-18 DB reversed-phase column	0.05 M phosphate buffer (pH 3)-acetonitrile	1	UV at 230	57
Shandon Hypersil BDS C18 column (5 μm particle size; 250 x 4.6 mm)	Methanol-citrate (0.08 M)-phosphate (0.04M) buffer (pH 3) (68 : 32)	1	240	67
Ultrasphere OSD column (250 x 4.6 mm)	Methanol-water (70:30)		298	68
C18 reversed-phase column	Phosphate buffer (pH 5.5)-methanol-acetonitrile (50 : 20 : 30)	1.4	230	69

4.6.3.1 HPLC Determination in Biological Matrices

Sensitive and selective HPLC methods that quantitate Nimesulide and its major metabolites in a variety of biological matrices have been reported. The matrices include human plasma [62, 66, 67, 69] and urine [57, 62].

4.6.3.2 HPLC Determination in Dosage Forms

A HPLC method useful for the separation and quantification of Nimesulide and its related compounds in drug raw material has been reported [63]. HPLC determination of Nimesulide in tablets by electrochemical detection has been described, with a mean recovery 101.07% and RSD equal to 1.44% [65]. Nimesulide has been determined in gels using this technique, with a mean recovery of 100.1% and RSD equal to 1.3% [68].

4.6.4 High Performance Thin-Layer Chromatography

High performance thin-layer chromatography (HPTLC) has been found to be useful for the determination of Nimesulide in biological matrices and in dosage forms.

4.6.4.1 HPTLC Determination in Biological Matrices

A rapid and sensitive HPTLC assay has been developed for the measurement of Nimesulide in human plasma, and has proved to be useful for pharmacokinetic studies [4]. The method includes a single-stage extraction procedure without the use of an internal standard. Known amounts of extract and Nimesulide (100 and 200 ng, as external standard) were spotted on a pre-coated silica gel 60 F_{254} plates by a Camag Linomat IV autosampler. Quantification was achieved using a Camag TLC scanner-3. The recovery of the method was 97%.

A second method involves use of a reversed- phase HPTLC on C18 bonded plates and has been reported to be a useful alternative to normal-phase TLC separations on silica gel [70].

4.6.4.2 HPTLC Determination in Dosage Forms

A simple, fast and accurate HPTLC method has been described for the determination of Nimesulide in pharmaceutical preparations [71]. Triclosan was used as an internal standard. The analysis involves use of Merck silica gel plates 60 F_{254}, a mobile phase of 9:1 chloroform / toluene, and a scanner detection wavelength of 300 nm. The average recovery was reported to be within pharmacopocial limits.

Another method reported [72] was based on the use of toluene-methanol (8:2) as the mobile phase and densitometry at 324 nm. Plots of peak area against the drug concentration were linear over the range of 20 - 200 ng/μL. The recovery was close to 100%.

4.7 Capillary Zone Electrophoresis

Capillary zone electrophoresis has been described as one of the methods for the determination of pKa of non-steroidal anti-inflammatory drugs which are sparingly soluble in aqueous solutions. The methodology relies on the measurement of the effective mobility of ionic species in electrophoretic buffers prepared at different pH values [42].

5. Stability

The stability of Nimesulide and its main metabolite (hydroxy-Nimesulide) has been reported in plasma [21]. The study indicated that blank plasma samples were spiked at 50 and 2500 ng/mL with Nimesulide and hydroxy-Nimesulide and stored frozen at -20° C. After ten months five samples/concentrations were thawed and analyzed in parallel with freshly spiked plasma samples. The respective recoveries found at two concentrations were 102.0 and 103.0% for Nimesulide and 100.0 and 101.0% for hydroxy-Nimesulide, confirming that no degradation of either drug occurred.

6. Metabolism

Nimesulide is mainly eliminated by metabolic transformation, with the major metabolite being 4-hydroxy-Nimesulide, or 4-nitro-2-(4′-hydroxy-phenoxy)methanesulfonanilide (compound **a** in Figure 16) [70, 73, 74]. Minor metabolites have also been detected. A unique metabolite of Nimesulide characterized as 4-amino-2-phenoxymethanesulfonanilide (compound **b**, Figure 16) resulting from reduction of the nitro group, has been reported in equine blood and urine samples of [75].

Recently Carini *et al.* [57] have characterized and quantitatively determined the main urinary metabolites (Figure 16) of Nimesulide in man following a single oral administration of 200 mg. These were found resulting from hydroxylation of the phenoxy nucleus (a), reduction of the nitro group (b), concomitant hydroxylation and reduction (c) and *N*-acetylation of b (d) and c (e) metabolites. The bulk of the metabolites were in conjugated form, and accounted for approximately 40% of the administered dose.

Electron impact ionization mass spectrometry was employed to elucidate the structures of the metabolites of Nimesulide [57].

7. Pharmacokinetics and Bioavailability

7.1 Pharmacokinetics

The pharmacokinetics of Nimesulide has been extensively investigated [76-84]. Studies have been carried out following oral or rectal administration in healthy volunteers, pediatric patients, patients with predisposition for altered pharmacokinetics, and in the elderly [74]. Bernareggi [81] has carried out a detailed clinical pharmacokinetic study of Nimesulide, and has reported that oral administration of Nimesulide tablet, granule, or suspension form in healthy human volunteers resulted in a rapid and extensive absorption of the drug. The mean peak concentration of 2.86 to 6.50 mg/L was reached within 1.22 to 2.75 hours of administration. The presence of food did not reduce either the rate or extent of absorption. Nimesulide is rapidly distributed and has an apparent volume of distribution ranging between 0.18 and 0.39 L/kg. The established mean terminal elimination half-life varied from 1.80 to 4.73 hours.

Figure 16. Metabolites of Nimesulide.

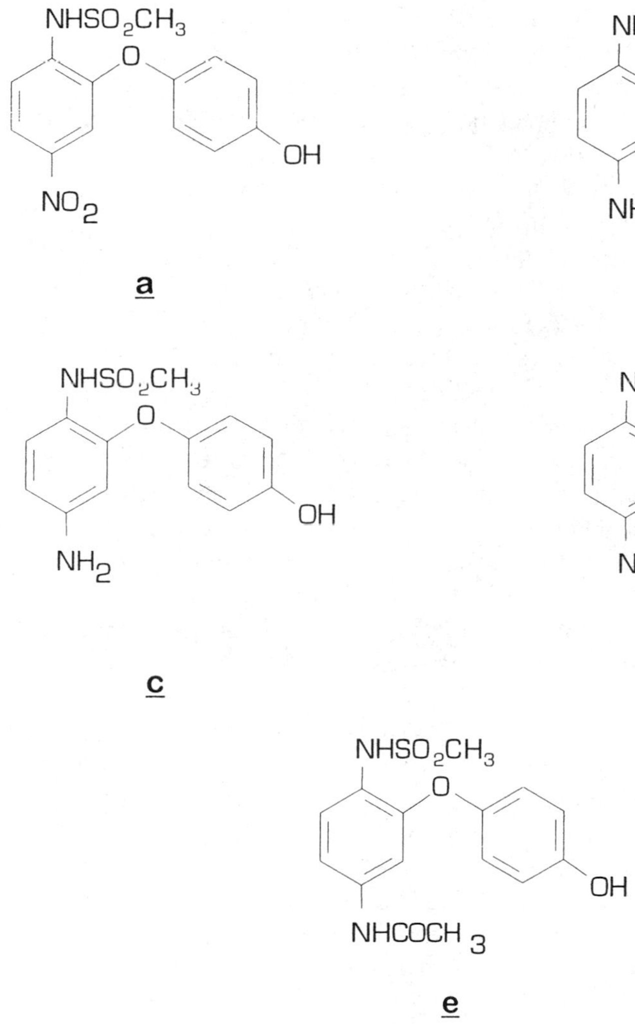

$NHSO_2CH_3$, O, OH, NO_2

a

$NHSO_2CH_3$, O, NH_2

b

$NHSO_2CH_3$, O, OH, NH_2

c

$NHSO_2CH_3$, O, $NHCOCH_3$

d

$NHSO_2CH_3$, O, OH, $NHCOCH_3$

e

Excretion of the unchanged drug in urine and feces is reported to be negligible. It is largely eliminated via its metabolites 50.5 to 62.5% in the urine and 17.9 to 36.2% in feces.

The total plasma clearance of Nimesulide is reported [81] to be 31.02 to 106.16 mL/h/kg, reflecting almost exclusive metabolic clearance. The pharmacokinetic profile of the drug in children and the elderly is not different from that of young healthy individuals. However, hepatic insufficiency is reported to remarkably reduce the rate of elimination of the drug necessitating a dose reduction (4 to 5 times) in patients with hepatic impairment. Moderate renal failure does not alter the pharmacokinetic profile of the drug.

Pharmacokinetic interaction between Nimesulide and other drugs given in combination such as glibenclamide, cimetidine, furosemide, theophylline, warfarin, digoxin, and antacids were reported to be absent or of no apparent clinical relevance. In another study Gandini *et al.* [76] have carried out first dose and steady state pharmacokinetics of Nimesulide and its 4-hydroxy metabolite in healthy volunteers. After a single dose of 200 mg, peak plasma concentration of Nimesulide (9.85 μg/mL) were reached at 3.17 hours and the half-life during the elimination phase was 4.95 hours. Plasma concentration on the seventh day, predicted from the results of the first day, were similar to the measured values. The study pointed out that pharmacokinetics of Nimesulide or its metabolite after single or repeated dose were not time or dose dependent.

Sengupta *et al.* [80] studied the analgesic efficacy and pharmacokinetics of topical Nimesulide gel in healthy human volunteers and carried out a double-blind comparison with piroxicam, diclofenac, and placebo. Nimesulide exhibited better efficacy than did diclofenac, piroxicam, and placebo. The superior analgesic activity of Nimesulide as gel formulation correlated with its pharmacokinetic profile. The study concluded that the topical route of administration may be a safe and effective alternative to the presently used oral or related routes [80].

Study of the pharmacokinetic profile of a parenteral formulation of Nimesulide demonstrated that Nimesulide intramuscularly administered may be superior to other routes of administration when fast onset of action is required [82]. A summarized version of the pharmacokinetic profile of Nimesulide in different dosage forms has been presented by Singla and coworkers [74]. In a review, Rainsford [83] analyzed the relationship of Nimesulide safety to its pharmacokinetics and concluded that Nimesulide is associated with a relatively low occurrence of adverse drug reactions, especially in the gastrointestinal tract. The reactions in the liver are within or below the general incidence with other NSAIDs'.

7.2 Bioavailability

The bioavailability of Nimesulide has been studied in healthy volunteers [78, 79, 81, 85]. The relative bioavailability of domestic or imported Nimesulide was determined after giving a single oral dose of 200 mg to ten volunteers in randomized crossover study by Chinese scientists Li *et al.* [79]. The plasma concentration of Nimesulide was assayed by high performance liquid chromatography. The concentration time curve of Nimesulide conformed to a one-compartment model, and the main parameter of domestic Nimesulide were:

$T_{½KC}$ = 3.61 ± 1.43 hours
T peak = 2.07 ± 0.63 hours
C_{max} = 0.46 ± 2.06 mg/L
AUC = 76.39 ± 17.62 mg/L/ hours

The relative bioavailability of domestic tablets was 92.2%. The results of these factors, analysis of variance, and Bayesian method, showed the two formulations to be bioequivalent.

In another study, the bioavailability of Nimesulide tablets and granules were identical, as shown in a random crossover design in humans [85]. Bernareggi [81] has reported that when Nimesulide was administered in suppository form, the C_{max} was lower and occurred later than after oral administration. The bioavailability of Nimesulide via suppository administration ranged from 54 to 64% relative to that of orally administered formulations.

The bioequivalence of two new pharmaceutical formulations of Nimesulide (sachet and effervescent tablets) with 100 mg Nimesulide, was investigated in 12 healthy volunteers [86]. No significant differences were found in the rate of bioequivalence analysis (C_{max} and T_{max}) and extent of Nimesulide absorption in the test (effervescent tablets) and in the reference drug (sachets).

8. Protein Binding

Bree *et al.* [73] have carried out a detailed equilibrium dialysis study of the binding of Nimesulide within human serum to isolated proteins and to erythrocytes. Within the range of therapeutic concentrations, Nimesulide was 99% bound to serum involving a non-saturated process ($NK_A = 91$). This binding was almost identical to binding of Nimesulide to serum albumin ($NK_A = 95$). Binding of Nimesulide to serum albumin was not affected by physiological concentrations of free fatty acids. The retention of Nimesulide by erythrocytes suspended in buffer was moderate (67%), although in whole blood no erythrocyte binding was observed because of the greater affinity of this drug for serum. Over the range of therapeutic concentrations (2.5 to 63 μmol/L) the free fraction of Nimesulide in serum remains constant.

Serum binding was decreased in samples obtained from patients with renal failure or hepatic cirrhosis associated with hypoalbuminemia and hyperbilirubinemia, respectively. The binding of Nimesulide at therapeutic concentrations was unaffected by warfarin, cefoperazone, furosemide, glibenclamide, tamoxifen, or digitoxin. However, valproic acid [73], fenofibrate [73, 87] (80 μmol/L), salicylic acid [87], tolbutamide [87] may displace Nimesulide on concurrent administration. It was reported that the principal metabolite of Nimesulide 4-hydroxy-Nimesulide, significantly increased the free fraction of the drug. Although methotrexate had no effect on the free fraction of Nimesulide, the free fraction of methotrexate was significantly increased in the presence of Nimesulide. It was also demonstrated by the study that there are two distinct Nimesulide binding sites, site I and site II, on serum albumin (10 μmol/L) with different affinities: site II $K_A = 3.57 \times 10^5$ L/mol and site I $K_A = 1.24 \times 10^5$ L/mol. It was indicated that Nimesulide binds to site II with higher affinity and to a lesser extent to site I. Bernareggi [81] has also reported that Nimesulide is extensively bound to albumin; the unbound fraction in plasma being 1%. The unbound fraction increased to 2 and 4% in patients with renal or hepatic insufficiency.

9. Toxicity

Nimesulide is the leading molecule of a new class of sulfonanilides among non-steroidal anti-inflammatory drugs that has shown a significant inhibitory selectivity towards cyclooxygenase-2 without affecting cyclooxygenase-1. This results in equivalent efficacy against pain and inflammation but with a better safety profile [88]. Nimesulide appears to be particularly useful for patients who have allergic hypersensitivity to aspirin or NSAIDs [89]. Studies have suggested Nimesulide as an alternative treatment in NSAIDs intolerant patients [90, 91].

There are very few reports of toxicity or adverse effects of Nimesulide. Though well-documented cases of acute hepatitis have not yet been reported with this drug, there is one report [92] on six patients who developed acute liver damage after initiation of Nimesulide. From clinical and histological data, it appears that both immunological and metabolic idiosyncratic reactions can be invoked as pathogenic mechanism of Nimesulide-induced liver disease. Although thrombocytopenia is a common feature in patients infected with HIV, one group of workers considered that thrombocytopenia in one of their patients was related to the use of Nimesulide [93].

A study has provided evidence that atopy and history of allergic reactions to antimicrobial drugs increase the likelihood of intolerance of Nimesulide in subjects allergic to NSAID's [94]. Risk factors for Nimesulide intolerance in patients with NSAID-induced skin disorders have been investigated [95]. Drug interactions with Nimesulide have been reviewed [87].

10. Dosage and Pharmaceutical Formulations

The usual adult oral and rectal dosage of Nimesulide for the treatment of a wide variety of inflammatory and pain states are 100 and 200 mg twice daily [84]. Nimesulide suspension and granules were commonly administered at a dosage of 5 mg/kg/day divided in two or three daily doses in paediatric clinical trials. Nimesulide has been formulated into various pharmaceutical forms. They include tablets [77, 79, 85], effervescent tablets [86], granules [77, 85], sachet [86], suspensions [77, 96], emulsions [97, 98], injectables [99 - 101] and suppositories [77].

Three multiple w/o/w emulsions containing Nimesulide when compared with drug suspension showed a slow and controlled release of Nimesulide with prolonged anti-inflammatory activity in rats [97]. Effect of the nature of oil, presence of electrolytes, phase volume ratios, and pH of the aqueous phase on the *in vitro* release of Nimesulide from multiple w/o/w emulsions was also studied [98]. Improved oral formulations for better drug delivery have been patented, such as compositions containing inclusion complexes for increased bioavailability [102], and granulcs for thc formation of rapidly disintegrating orally soluble tablets [103]. These tablets are capable of disintegrating inside the mouth of the patient within a very short time.

A pharmaceutical preparation comprising coated capsules or tablets containing liposome powder encapsulating Nimesulide has been patented for improved oral bioavailability [104]. It is reported that blood levels of Nimesulide in volunteers was 7.31 μg/mL as compared with 2.69 μg/mL. A composition containing piperine, a bioavailability enhancer, has also been patented which is characterized in having clinically significant increased bioavailability when compared to the known formulations [105]. Recently, novel drug delivery approaches have been applied and Nimesulide has been formulated as solid dispersion [106], osmotic pumps [107], and transdermal delivery systems [108, 109].

A number of pharmaceutical preparations containing Nimesulide for topical use have been developed. These include gels [110-112], creams [113, 114], and liquid crystals [115]. A hydroalcoholic mouthwash containing Nimesulide has been developed for local use, with application to the oral and rhinopharyngeal cavity for the treatment of inflammation of oral and rhinopharyngeal mucosa [116].

Acknowledgement

The authors would like to thank Mr. Vivek Kumar Sharma for technical assistance in the preparation of the manuscript.

References

1. ***Martindale: The Complete Drug Reference***, 32nd edn., The Pharmaceutical Press, London, 1999, p 63.

2. S. Budavari, ed., ***The Merck Index***, 12th edn., Merck & Co., White House Station, N.J., 1996, p 6640.

3. **Pharmeuropa**, Vol 10, No 4, Council of Europe, France, 1998, p 608.

4. K.K. Pandya, M.C. Satia, M.C. Modi, I.A. Modi, R.I. Modi, B.K. Chakravarthy, and T.P. Gandhi, *J. Pharm Pharmacol.*, **49**, 773 (1997).

5. G.G.I. Moore and J.K. Harrington, **U.S. Patent 3,840,597** (1974).

6. A. Singh and A. Prasad, unpublished data.

7. S.Chen and C. Luo, *Huaxi Yaoxue Zazhi*, **13**, 180 (1998); *Chem Abs.*, **130**, 196478 g (1999).

8. G. Cignarella, P. Vianello, F. Berti and G. Rossoni, *Eur. J. Med. Chem.*, **31**, 359 (1996).

9. Y. Konishi, A. Denta and E. Okashima, **Japan Patent JP 11 92,366 [99 92, 366]**; *Chem. Abs.*, **130**, 291579z (1999).

10. Y. Oyanagi, **Japan Patent JP 04,124,130 [92, 124, 130]**; *Chem. Abs.*, **117**, 143464 z (1992).

11. Y Yoshikawa, Y. Ochi, K. Sekiuchi, H. Saito, and K. Hatayama, **Japan Patent JP 02 22,260 [90 22, 260]**; *Chem. Abs.*, **113**, 77922m (1990).

12. P. Del Soldato, F. Sannicolo and T. Benincori, **PCT Int. Appl. WO 96 15,809**; *Chem. Abs.*, **125**, 114476q (1996).

13. D.P.Carr, R. Henn, J.R. Green, and I. Boettcher, *Agents Actions*, **19**, 374 (1986).

14. M. Sato, M. Hizaki, Y. Tada, and M. Yamada, **Japan Patent JP 09,202,728 [97,202,728]**; *Chem. Abs.*, **127**, 210348m (1997).

15. P. Del Soldato and F. Sannicolo, **PCT Int. Appl. WO 98 09,948**; *Chem. Abs.*, **128**, 217188p (1998).

16. K. Yoshikawa, M. Hasegawa, M. Suzuki, Y. Shimazaki, M. Ohtani, S. Saito, and M. Goi, **PCT Int. Appl. WO 97 46,520**; *Chem. Abs.*, **128**, 61353a (1998).

17. K.M. Lundy and A.P. Ricketts, **PCT Int. Appl. WO 98 50,033**, *Chem. Abs.*, **130**, 10625z (1999).

18. Database, Panacea Biotec Ltd., Lalru, India.

19. L. Dupont, B. Pirotte, B. Masereel, and J. Delarge, *Acta Crystallogr, Sect C; Cryst. Struct. Commun.*, **C51**, 507 (1995).

20. A. Kapoor, D.K. Majumdar, and M.R. Yadav, *Indian J. Chem. Sect. B*, **37B**, 572 (1998).

21. C. Giachetti and A. Tenconi, *Biomed Chromatogr.*, **12**, 50 (1998).

22. G. Piel, B. Pirotte, I. Delneuville, P. Neven, G. Llabres, J. Delarge, and L Delattre, *J Pharma Sci.*; **86**, 475 (1997)

23. I Sicart Girona, **Ger. Offen. D.E. 4,116,659**; *Chem. Abs.*, **116**, 86185n (1992)

24. I Sicart Girona, **Fr. Demande FR 2,662,360**; *Chem. Abs.*, **117**, 14441r (1992).

25. G. Maffione, **PCT Int. Appl. WO 94 02,177**; *Chem. Abs.*, **120**, 200446h (1994).

26. J Geczy, **PCT Int. Appl. WO 94 28,031**; *Chem. Abs.*, **122**, 196975b (1995)

27. F. Barbato, B. Cappello, M.I. La Rotonda, A Miro, and P. Morrica, *Proc. 24th Int. Syrup. Controlled Release Bioact. Mater*, 385 (1997).

28. P.R. Vavia and N.A. Adhage, *Drug Dev. Ind. Pharm.*, **25**, 543 (1999).

29. G. Scolari, F. Lazzarin, C. Fornaseri, V. Carbone, S. Rengo, M. Amato, D. Cicciu, D. Braione, S. Argentino, A. Morgantini, C. Bassetti, M. Tramer, and G.C. Monza, *Int. J. Clin. Pract.*, **53**, 345 (1999).

30. G. Piel, I. Delneuville, and L. Delattre, *Proc. 8th Int. Syrup. Cyclodextrins*, 487 (1996).

31. B. Pirotte, G. Piel, P. Neven. I. Delneuville, and J. Geczy, **PCT Int. Appl. WO 95 34,533**; *Chem. Abs.*, **124**, 242309d (1996).

32. K.P.R. Chowdary, A.R. Rani, and L.S. Latha, *East Pharm.*, **41**, 163 (1998).

33. K.P.R. Chowdary, Y.S. Devi, and M. Deepika, *East Pharm.*, **42**, 119 (1999).

34. V. De Tommaso, **PCT Int. Appl. WO 99 41,233**; *Chem. Abs.*, **131**, 161641y (1999).

35. T. Monti and W. Mossi, **Eur. Pat. Appl. EP 937,709**; *Chem. Abs.*, **131**, 157646m (1999).

36. M.G. Di Schiena and I.M. Di Schiena, **Eur. Pat. Appl. EP 869,117**; *Chem. Abs.*, **129**, 275702h (1998).

37. S.J. Macnaughton, I. Kikic, N.R. Foster, P. Alessi, A. Cortesi, and I. Colombo, *J. Chem. Eng. Data*, **41**, 1083 (1996).

38. C Hansch, P.G. Sammes, and J.B. Taylor, **Comprehensive Medicinal Chemistry**, vol 6, Pergamon, Oxford, 1990, p711.

39. P.R.B. Fallavena and E.E.S. Schapoval, *Int. J. Pharm.*, **158**, 109 (1997).

40. E. Magni, **Drug Invest. (Suppl.), 3**, 1 (1991).

41. S.Singh, N Sharda, and L. Mahajan, *Int. J. Pharm.*, **176**, 261 (1999).

42. N. Chauret., D.K. Lloyd, D. Levorse, and D.A. Nicoll-Griffith, *Pharm. Sci.*, **1**, 59 (1995).

43. A. Alvarez - Lueje, P. Vasquez, L.J. Nunez-Vergara, and J.A. Squella, *Electroanalysis*, **9**, 1209 (1997).

44. J.A. Squella, P. Gonzalez, S. Bollo and L.J. Nunez-Vergara, *Pharm, Res.*, **16**, 161 (1999).

45. Y.V.R. Reddy, P.R.K. Reddy, C.S. Reddy, and S.J. Reddy, *Indian J. Pharm, Sci.*, **58**, 96 (1996).

46. C.S.R. Lakshmi, M.N. Reddy, and P.Y. Naidu, *Indian Drugs*, **35**, 519 (1998).

47. P.R.B. Fallavena and E.V.S. Schapoval, *Rev. Bras. Farm*, **76**, 30 (1995); *Chem. Abs.*, **124**, 15602a (1996)

48. S.J. Rajput and G. Randive, *East Pharm.*, **40**, 113 (1997).

49. O.S. Kamalapurkar and Y. Harikrishna, *East Pharm.*, **40**, 145 (1997).

50. K.P.R. Chowdary, G.D. Rao, and I.S. Babu, *Indian Drugs*, **34**, 396 (1997).

51. M.N. Reddy, K.S. Reddy, D.G. Shankar, and K. Sreedhar, *Indian J. Pharm. Sci.*, **60**, 172 (1998).

52. M.N. Reddy, K.S. Reddy, D.G. Sankar, and K. Sreedhar, *East Pharm.*, **41**, 119 (1998).

53. C.S.R. Lakshmi and M.N. Reddy, *J. Inst. Chem (India)*, **70**, 151 (1998).

54. K.P.R Chowdary, K.G. Kumar, and G.D. Rao, *Indian Drugs*, **36**, 185 (1999).

55. K.E.V. Nagoji, S.S. Rao, M.E.B. Rao, K.V.K. Rao, *East Pharm.*, **42**, 117 (1999).

56. A. Manna, I. Ghosh, L.K. Ghosh, and B.K. Gupta, *East Pharm.*, **42**, 101 (1999).

57. M. Carini, G. Aldini, R. Stefani, C. Marinello, and R.M. Facino, *J. Pharm Biomed Anal.*, **18**, 201 (1998).

58. J.M. Young, S. Panah, C. Satchawatcharaphong, and P.S. Cheung, *Inflammation Res.*, **45**, 246 (1996).

59. A.A. Shah, F.E. Murray, and D.J. Fitzgerald, *Rheumatology (Oxford)*, **38 (suppl. 1)**,19, (1999).

60. F. Giuliano and T.D. Warner, *Br. J. Pharmacol.*, **126**, 1824 (1999).

61. S. G. Navalgund, D.H. Khanolkar, P.S. Prabhu, P.S. Sahasrabudhe, and R.T. Sane, *Indian Drugs*, **36**, 173 (1999).

62. D. Castoldi, V. Monzani, and O. Tofanetti, *J. Chromatogr*, **425**, 413 (1988).

63. A. Nonzioli, G. Luque, and C. Fernandez, *J. High Resolut Chromatogr.*, **12**, 413 (1989)

64. Z. Zeng and H. Zhang, *Zhongguo Yaoxue Zazhi (Beijing)*, **31**, 610 (1996); *Chem. Abs.*, **126**, 242955c (1997).

65. A. Alvarez-Lueja, P.Vasquez, L.J.. Nunez-Vergara, and J.A. Squella, *Anal. Lett.*, **31**, 1173 (1998).

66. L. Ding, Z. Zhang, L. Gao, H. Chen, G. Yu, and D. An, *Zhongguo Yaoke Daxue Xuebao*, **29**, 370 (1998), *Chem. Abs.*, **130**, 177049 b (1999).

67. D.J. Jaworowicz Jr., M.T. Filipowski, and K.M.K. Boje, *J. Chromatogr. B: Biomed Sci. Appl.*, **723**, 293 (1999).

68. X. Pan, F. Ma, and J. Lu, *Zhongguo Yiyuan Yaoxue Zazhi*, **19**, 207 (1999); *Chem. Abs.*, **131**, 9723u (1999).

69. G. Khaksa and N. Udupa, *J. Chromatogr., B: Biomed Sci. Appl.*, **727**, 241 (1999).

70. G.L. Swaisland, L.A. Wilson, and I.D. Wilson, *J. Planar Chromatogr.*, **10**, 372 (1997).

71. A. Sacahan and P. Trivedi, *Indian Drugs*, **35**, 762 (1998).

72. R.T. Sane, S.G. Joshi, M. Francis, A.R. Khatri, and P.S. Hijli, *J. Planar Chromatogr.*, **12**, 158 (1999).

73. F.Bree. P.Nguyen, S. Urien, E. Albengres, A. Macciocchi, and J.P. Tillement, *Drugs*, **46 (suppl. 1)**, 83 (1993).

74. A. K. Singla, M Chawla, and A. Singh, *J. Pharm. Pharmacol.*, **52**, 467 (2000).

75. P. Sarkar, J.M. McIntosh, R. Leavitt, and H. Gouthro, *J. Anal. Toxicol.*, **21**, 197 (1997).

76. R. Gandini, C. Montalto, D. Castoldi, V. Monzani, M.L. Nava, I Scaricabarozzi, G. Vargiu, and I. Bartosek, *Farmaco*, **46**, 1071 (1991).

77. A Bernareggi, *Drugs*, **46 (suppl 1)**, 64 (1993).

78. J.Li, X.Chen, Y.Zhang, Y.Jin, C. Chen, and S. Xu, *Zhongguo Linchuang Yaoli Zazhi*, **10**, 106 (1994); *Chem. Abs.*, **122**, 45708r (1995).

79. J.Li, X.Tang, Y.Jin, Y.Zhang, and S.Xu, *Zhongguo Yaolixue Tongbao*, **12**, 185 (1996); *Chem. Abs.*, **125**, 308887v (1996).

80. S. Sengupta, T. Velpandian, S.R. Kabir, and S.k. Gupta, *Eur. J. Clin Pharmacol.*, **54**, 541 (1998).

81. A Bernareggi, *Clin. Pharmacokinet.*, **35**, 247 (1998).

82. S.K. Gupta, R.K. Bhardwaj, P. Tyagi, S. Sengupta, and T. Velpandian, *Pharmacol. Res.*, **39**, 137 (1999).

83. K.D. Rainsford, *Rheumatology*, **38 (suppl. 1)**, 4 (1999).

84. R. Davis and R.N. Brogden, *Drugs*, **48**, 431 (1994).

85. A.Alessandrini, E.Ballarin, A.Bastianon, and C. Migliavacca, *Clin. Ter (Rome)*, **118**, 177 (1986).

86. S. Contos, A.S. Tripodi, and M. De Bernardi, *Acta Toxicol Therap.*, **17**, 339 (1996).

87. E. Perucca, *Drugs*, **46 (suppl. 1)**, 79 (1993).

88. X. Rabasseda, *Drugs Today*, **33**, 41 (1997).

89. P.A. Insel, in **Goodman & Gilman's The Pharmacological Basis of Therapeutics,** 9th edn., J.G. Hardman and L.E. Limbird, eds., McGraw Hill, New York, 1996, p. 644.

90. M.A. Morais de Almeida, A.P.Gaspar, F.S. Carvalho, J.M. Abreu Nogueira, and J.E. Rosado Pinto, *Allergy Asthma Proc.*, **18**, 313 (1997).

91. D.Quaratino, A.Romano, G.Papa, M. Di Fonso, F. Giuffreda, F.P. D'Ambrosio, and A.Venuti, *Ann.Allergy, Asthma, Immunol.*, **79**, 47 (1997).

92. W.Van Steenbergen, P. Peeters, J. De. Bondt, D. Staessen, H. Buscher, T Laporta, T. Roskams, and V. Desmet, *J. Hepatol.*, **29**, 135 (1998).

93. M.B. Pasticci, F. Menichetti, and F. DiCandilo, *Ann. Intern. Med.*, **112**, 233 (1990).

94. E.A. Pastorello, C. Zara, G.G. Riariosforza, V. Pravettoni, and C. Incorvaia, *Allergy (Copenhagen)*, **53**, 880 (1998).

95. R. Asero, *Ann. Allergy, Asthma, Immunol.*, **82**, 554 (1999).

96. M. Borsa. **Eur. Pat. Appl. EP 843,998**; *Chem. Abs.*, **129**, 32312b (1998).

97. P. Tandon and B. Mishra, *Acta Pharm. Turc.*, **41**, 15 (1999).

98. P. Tandon and B. Mishra, *Acta Pharm. Turc.*, **41**, 9 (1999).

99. A. Singh and R. Jain (Panacea Biotec Ltd.) **Japan Patent JP 11 228, 448 [99 228, 448]** *Chem. Abs.,* **131**, 175077f (1999)

100. R. Jain and A.Singh (Panacea Biotec Ltd.), **United States Patent 5,688,829**; *Chem. Abs.,* **128**, 26936w (1998).

101. R. Jain and A.Singh (Panacea Biotec Ltd.), **Canadian Patent Appl. CA2,202,425**. *Chem. Abs.,* **128**, 248573q (1998)

102. G. Santus, R. Golzi, C. Lazzarini and L. Marcelloni, **PCT Int. Appl. WO 99 47,172**; *Chem. Abs.,* **131**, 248243u (1999).

103. D. Bonadeo, F. Ciccarello and A. Pagano, **PCT Int. Appl. WO 99 04,758**; *Chem. Abs.,* **130**, 158406z (1999).

104. J. Garces Garces, A. Bonilla Munoz and A. Parente Duena, **Eur. Pat. Appl. EP 855,179**; *Chem. Abs.,* **129**, 166204b (1998).

105. A. Singh and R. Jain (Panacea Biotec Ltd.), **Eur. Pat. Appl. EP 935,964**; *Chem. Abs.,* **131**, 161660d (1999).

106. S.R.R. Hiremath, S. Suresh, S Praveen, A. Thomas, T. Sriramulu, G. Jaya, and M. Ruknuddin, *Indian J. Pharm Sci.*, **60**, 326 (1998).

107. R.K. Verma and B. Mishra, *Pharmazie*, **54**, 74 (1999).

108. R. Ilango, S. Kavimani, K.S. Kumar, K.R. Deepa, and B. Jaykar, *East Pharm.*, **41**, 123 (1998).

109. R. Jain and A.Singh (Panacea Biotec Ltd.), **United States Patent 5,716,609**; *Chem. Abs.,* **128**, 158954k (1998).

110. S. Bader, E. Hausermann and T. Monti, **PCT Int. Appl. WO 98 37,879**; *Chem. Abs.,* **129**, 207203 w (1998).

111. P.L. M. Giorgetti, **PCT Int. Appl. WO 98 01,124**; *Chem. Abs.,* **128**, 132436t (1998).

112. S.K. Gupta, J Prakash, L. Awor, S. Joshi, T. Velpandian, and S. Sengupta, *Inflammation Res.*, **45**, 590 (1996).

113. M.G. Di Schiena, **Eur. Pat. Appl. EP 880,965**; *Chem. Abs.,* **130**, 43327g (1999).

114. S. Miyata, Y. Taniguchi, K. Masuda and Y. Kawamura, **PCT Int. Appl. WO 96 11,002**; *Chem. Abs.,* **125**, 41817q (1996).

115. S. Bader, E. Hausermann, and T. Monti, **PCT Int. Appl. WO 98 37,868**; *Chem. Abs.,* **129**, 207224d (1998).

116. P.L.M. Giorgetti, **PCT Int. Appl. WO 98 47,501**; *Chem. Abs.,* **129**, 335774v (1998).

PROCAINAMIDE HYDROCHLORIDE

Mohammed S. Mian, Humeida A. El-Obeid,
and Abdullah A. Al-Badr

Department of Pharmaceutical Chemistry
College of Pharmacy
King Saud University
P.O. Box 2457
Riyadh-11451
Kingdom of Saudi Arabia

1075-6280/01 $35.00

Contents

1. Description

1.1 Nomenclature

1.1.1 Systematic Chemical Names

4-Amino-N-[2-(diethylamino)ethyl]benzamide mono-hydrochloride [1]

Benzamide, 4-amino-N-[2-(diethylamino)-ethyl]-mono-hydrochloride [2]

N-(*p*-Aminobenzoyl)-2-diethylaminoethylamine mono-hydrochloride

N-[2-(Diethylamino)ethyl]-*p*-aminobenzamide mono-hydrochloride

p-Aminobenzamide, N-[2-(diethylamino)ethyl] mono-hydrochloride

Benzoic acid amide, 4-amino N-[2-(diethylamino)ethyl] mono-hydrochloride.

p-Aminobenzoic acid amide, N-[2-(diethylamino)ethyl] mono-hydrochloride.

1.1.2 Nonproprietary Names

Procaine amide hydrochloride; Procainamide hydrochloride

1.1.3 Proprietary Names

Amidoprocain; Amisalin; Novocamid; Novocainamid; Procamide; Procainde amide hydrochloride; Procan-SR; Procanbid; Procapan; Procardyl; Promide; Pronestyl hydrochloride; Supicaine amide hydrochloride; Procainamide durules [1]

1.2 Formulae

1.2.1 Empirical Formulae, Molecular Weights, CAS Numbers [1]

Procainamide:	$C_{13}H_{21}N_3O$	(235.29)	[51-06-9]
Procainamide HCl:	$C_{13}H_{22}ClN_3O$	(271.79)	[614-39-1]

1.2.2 Structural Formula

1.2.2.1 Procainamide base

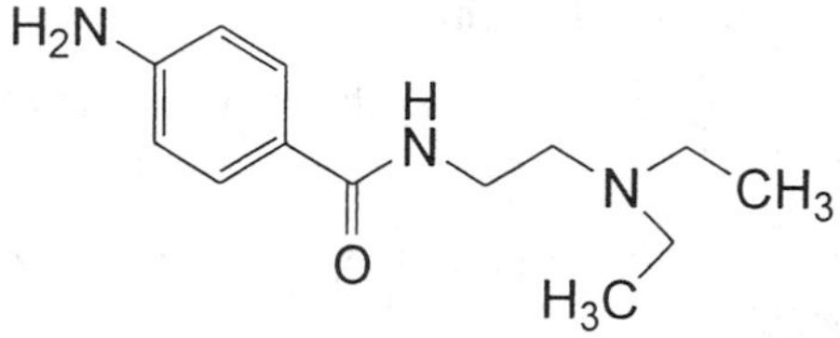

1.2.2.2 Procainamide hydrochloride

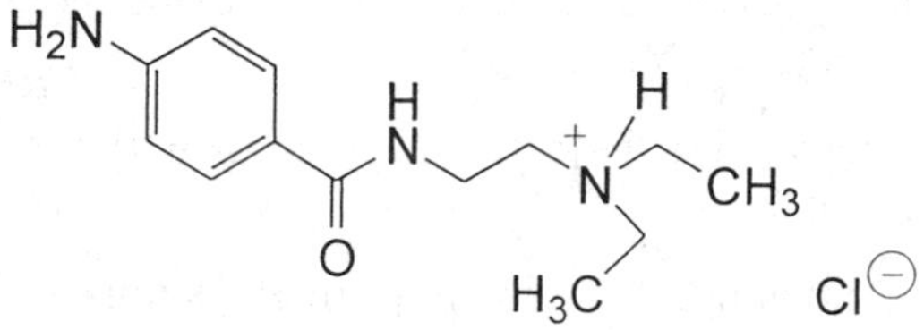

1.3 Elemental Analysis

The calculated elemental composition is as follows [1]:

	Procainamide	Procainamide HCl
carbon:	66.36%	57.45%
hydrogen:	8.99%	8.16%
oxygen:	6.79%	5.89%
nitrogen:	17.86%	15.46%
chlorine:	–	13.04%

1.4 Appearance

Procainamide hydrochloride is a white to yellowish-white crystalline powder, which is odorless and hygroscopic [3].

1.5 Uses and Applications

In 1935, Mautz [4] demonstrated that direct application of procaine (a compound that differs from Procainamide by replacement of the amide linkage by the ester structure), to the myocardium, elevated the threshold of ventricular muscle to electrical stimulation. Extension of this observation by numerous workers established that the cardiac actions of the local anesthetic resemble those of quinidine. However, the therapeutic value of procaine as an anti-fibrillatory and anti-arrhythmic agent is limited by its short duration of action, as a result of the rapid enzymatic hydrolysis of the drug, and by its prominent effects upon the central nervous system. A systematic study of the congeners and metabolites of procaine was undertaken to find a compound with clinically useful quinidine-like actions, and this led to the introduction of Procainamide as a cardiac drug [5]. The drug is useful for the treatment of a variety of arrhythmias, and it can be administered by several routes. Unfortunately, its potency and versatility are marred by its short duration of action and a high incidence of adverse reactions when it is chronically used [6].

2. Method(s) of Preparation

In their earlier analytical profile, Poet and Kadin [7] reported that *p*-nitrobenzoic acid is converted to the acid chloride by reaction with thionyl chloride. *p*-nitrobenzoyl chloride is then reacted with diethyl-aminoethylamine. The resultant nitro condensation product is treated with hydrochloric acid and reduced to yield the final product. This procedure is outlined below.

NO_2 / COOH + $SOCl_2$ → NO_2 / COCl → (H_2N–CH₂CH₂–N(Et)₂)

NO_2 / O=C–N(H)–CH₂CH₂–N(Et)₂ → (reduction / HCl) → NH_2 / O=C–N(H)–CH₂CH₂–$\overset{+}{N}$–H $Cl^{\ominus}$

The authors [7] also listed the other methods of preparation that have been described by various workers (references cited therein).

3. Physical Properties

3.1 Ionization Constant

The pKa pf Procainamide has been reported to be 9.2 (at 20°C) [9].

3.2 Solubility Characteristics

The hydrochloride salt of Procainamide is soluble at 20°C in less than 1 part of water, in 2 parts of alcohol, and in 140 parts of chloroform. The substance is also very slightly soluble in ether [8].

A 10% solution of Procainamide hydrochloride has a pH of 5 to 6.5 [8].

3.3 Partition Coefficient

Consistent with its hydrophilic nature, the value reported for log P in the ethyl acetate / pH 7.4 buffer system is –1.5 [9].

3.4 X-Ray Powder Diffraction

The X-ray powder diffraction pattern of Procainamide hydrochloride was determined using a Philips automated x-ray diffraction spectrogoniometer equipped with a PW 1730/10 generator. The radiation source was a copper high intensity x-ray tube (Cu anode, 2000 W, λ = 1.5418 Å), and operated at 70 kV and 35 mA. The monochromator was a curved single crystal (PW 1752), and the divergence and receiving slits were 1□ and 0.1□, respectively. The powder pattern was obtained at a scanning speed of 0.2 degrees 2-θ/sec. The goniometer was aligned using silicon before use.

The powder pattern of Procainamide hydrochloride is shown in Figure 1, and a table of derived parameters (scattering angles, d-spacings, and relative intensities) is given in Table 1.

Figure 1 X-ray powder pattern of Procainamide hydrochloride.

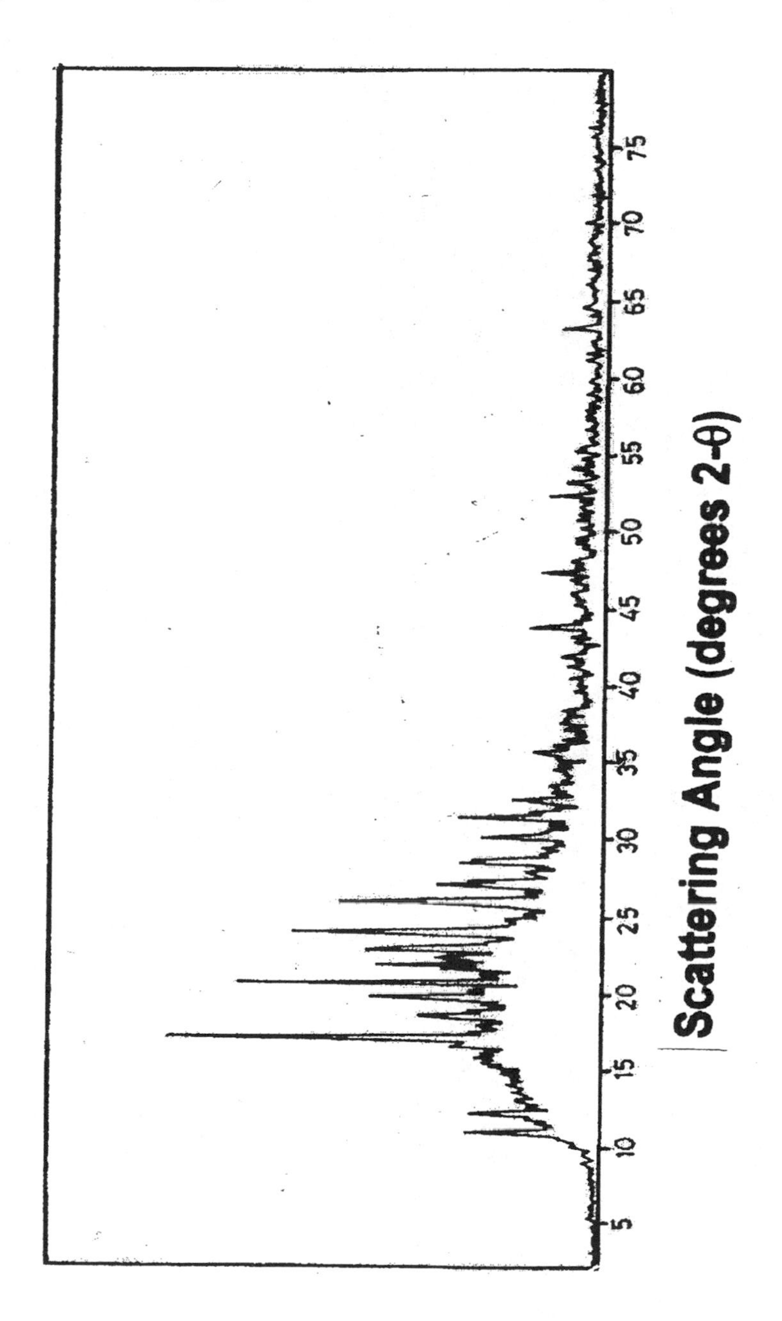

Table 1

Crystallographic Data from the X-ray Powder Diffraction Pattern of Procainamide Hydrochloride

Scattering Angle (degrees 2-θ)	d-spacing (Å)	Relative Intensity (%)
10.951	8.0725	29.65
12.176	7.2631	29.30
17.130	5.1719	100.00
18.567	4.7748	40.85
19.863	4.4661	52.51
20.794	4.2682	83.57
21.957	4.0447	51.20
22.429	3.9607	37.48
22.948	3.8723	53.18
24.058	3.6960	70.46
25.964	3.4289	60.06
27.076	3.2906	37.02
27.340	3.2593	30.02
28.527	3.1264	31.67
30.141	2.9626	26.61
31.390	2.8474	31.99
32.583	2.7458	20.26
35.642	2.5169	14.77
36.474	2.4614	8.87
43.856	2.0626	16.58
47.352	1.9182	13.51
52.249	1.7494	12.43
63.221	1.4696	9.13

Figure 2 Expanded x-ray powder pattern of Procainamide hydrochloride.

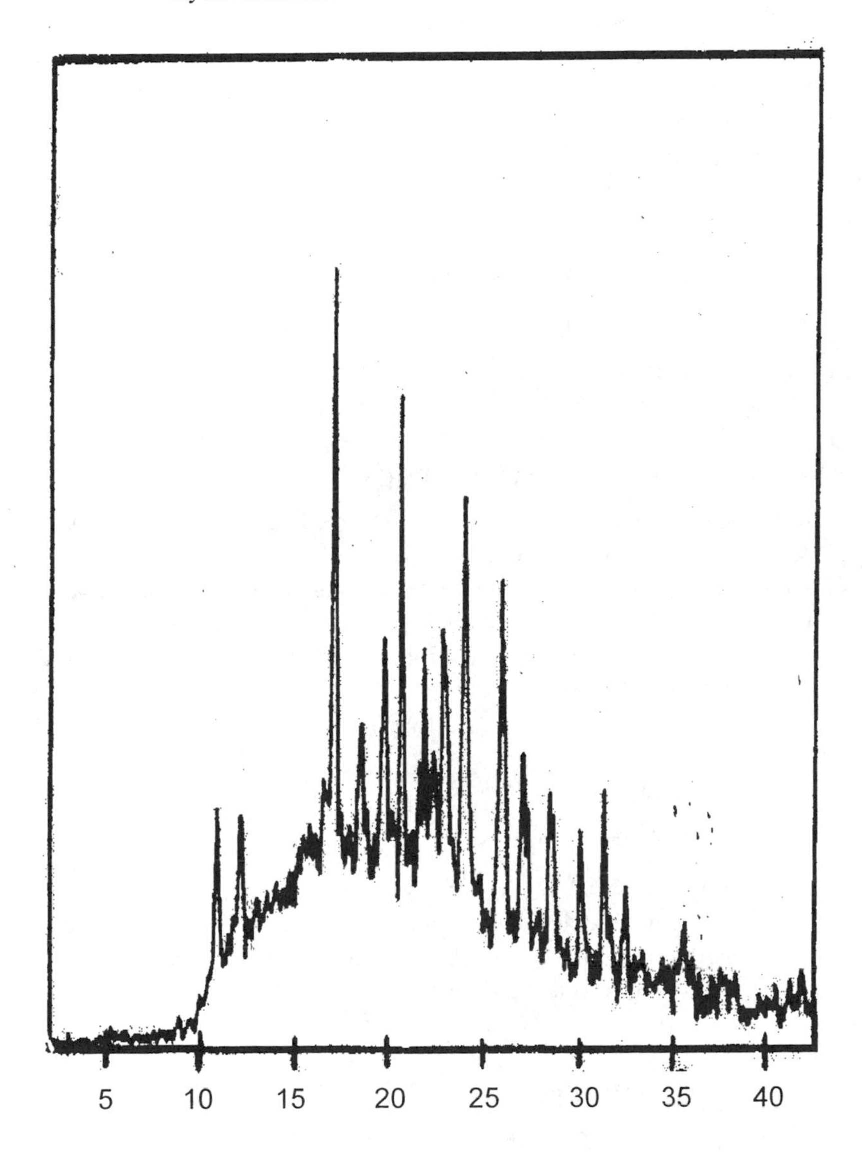

Owen *et al* presented an X-ray powder diffraction data for Procainamide and other local anesthetics [103]. The data for the drug are tabulated in term of lattice spacing (d in \AG/A) and the relative intensities of lines. In addition, a classification was provided of the three most intense lines for each of the local anesthetics, with their relative intensities arranged in descending numerical order with regard to d spacing of the most intense line from each pattern.

3.5 Thermal Methods of analysis

3.5.1 Melting Behavior

Procainamide hydrochloride is reported to melt over the range of 165-169°C [1].

3.5.2 Differential Scanning Calorimetry

Differential scanning calorimetry was performed using a Dupont DSC model TA 9900 Compiler/ thermal analyzer. The thermogram that is shown in Figure 3 was carried out over the range of 120 to 210°C at a heating rate of 10°C /min. The enthalpy of fusion for the sample was found to be 32.7 kJ/mole (7.81 Kcal/mole), and the purity of the sample was calculated as 99.33%.

3.6 Spectroscopy

3.6.1 UV/VIS Spectroscopy

Using a Shimadzu 1601 PC UV/VIS spectrophotometer, the UV spectrum of Procainamide hydrochloride was obtained in ethanol and in water over the wavelength region of 200 to 340 nm. The spectra are shown in Figure 4, and the following results were obtained:

Ethanol: λ_{max} = 292.1 nm
A(1%, 1 cm) = 609 molar absorptivity = 1.655×10^3

Water: λ_{max} = 278.6 nm
A(1%, 1 cm) = 609 molar absorptivity = 1.655×10^3

Clarke [9] reported an absorption maximum in aqueous alkali at 275 nm (A: 693a).

Figure 3. Differential scanning calorimetry thermogram of Procainamide hydrochloride.

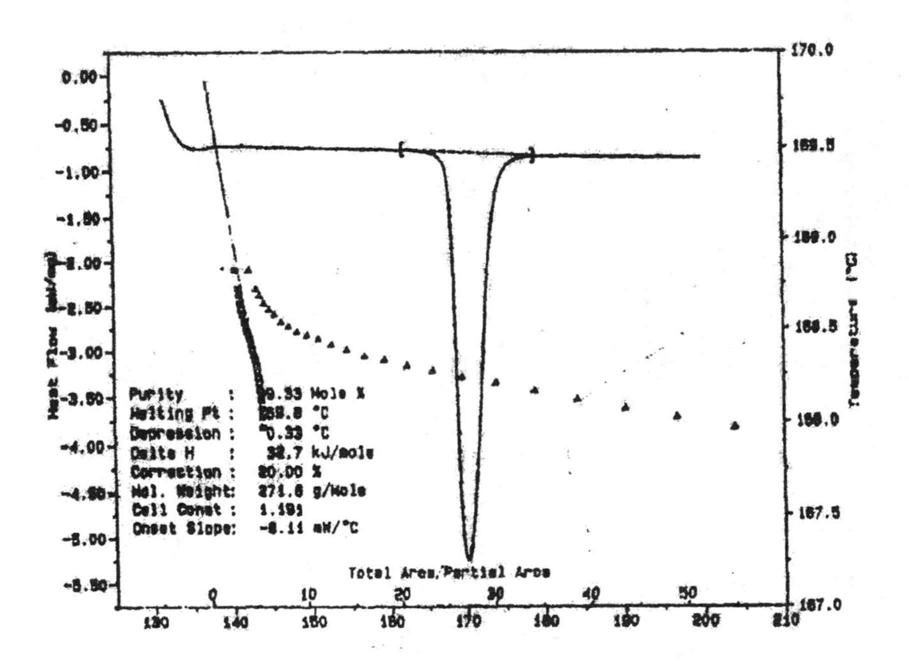

Temperature (°C)

Figure 4. Ultraviolet absorption spectrum of Procainamide hydrochloride in (a) ethanol, and in (b) water.

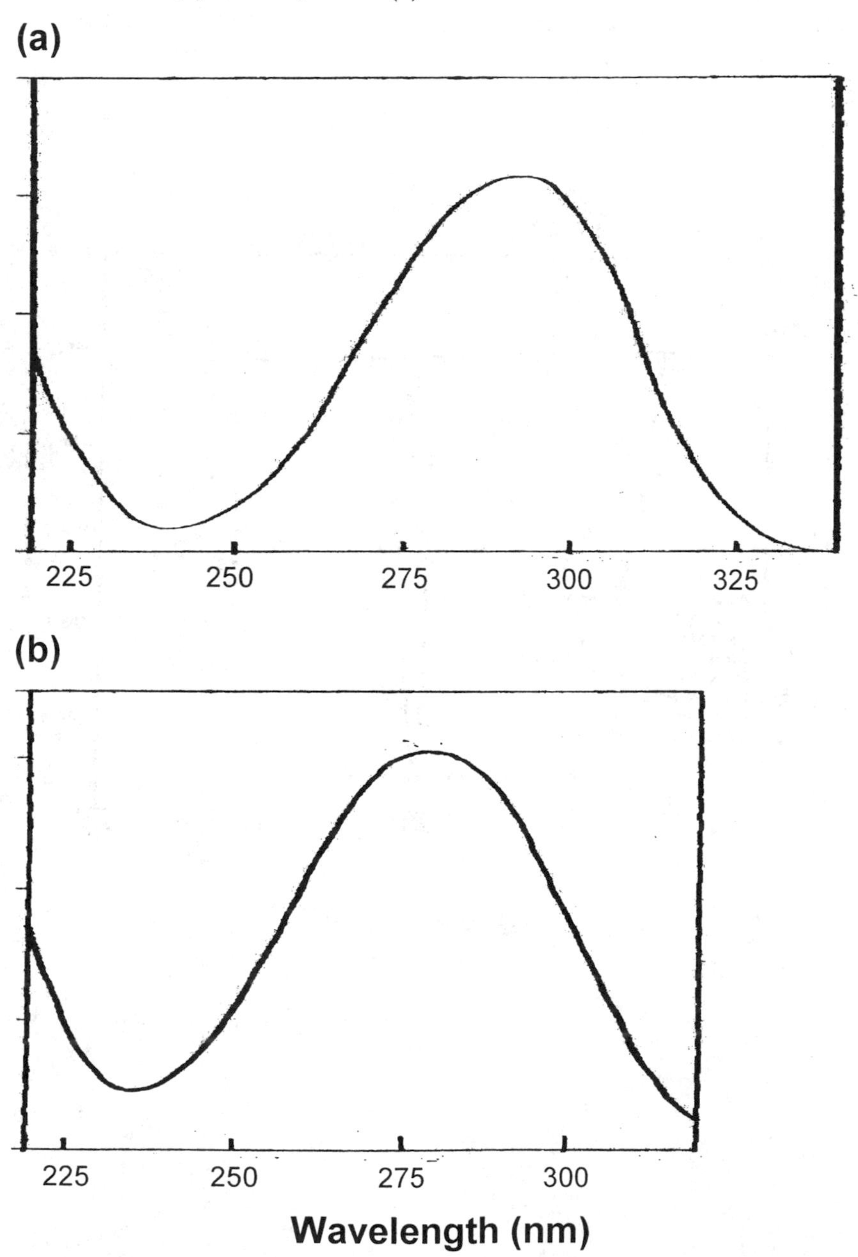

Figure 5. Infrared absorption spectrum of Procainamide hydrochloride in a KBr pellet.

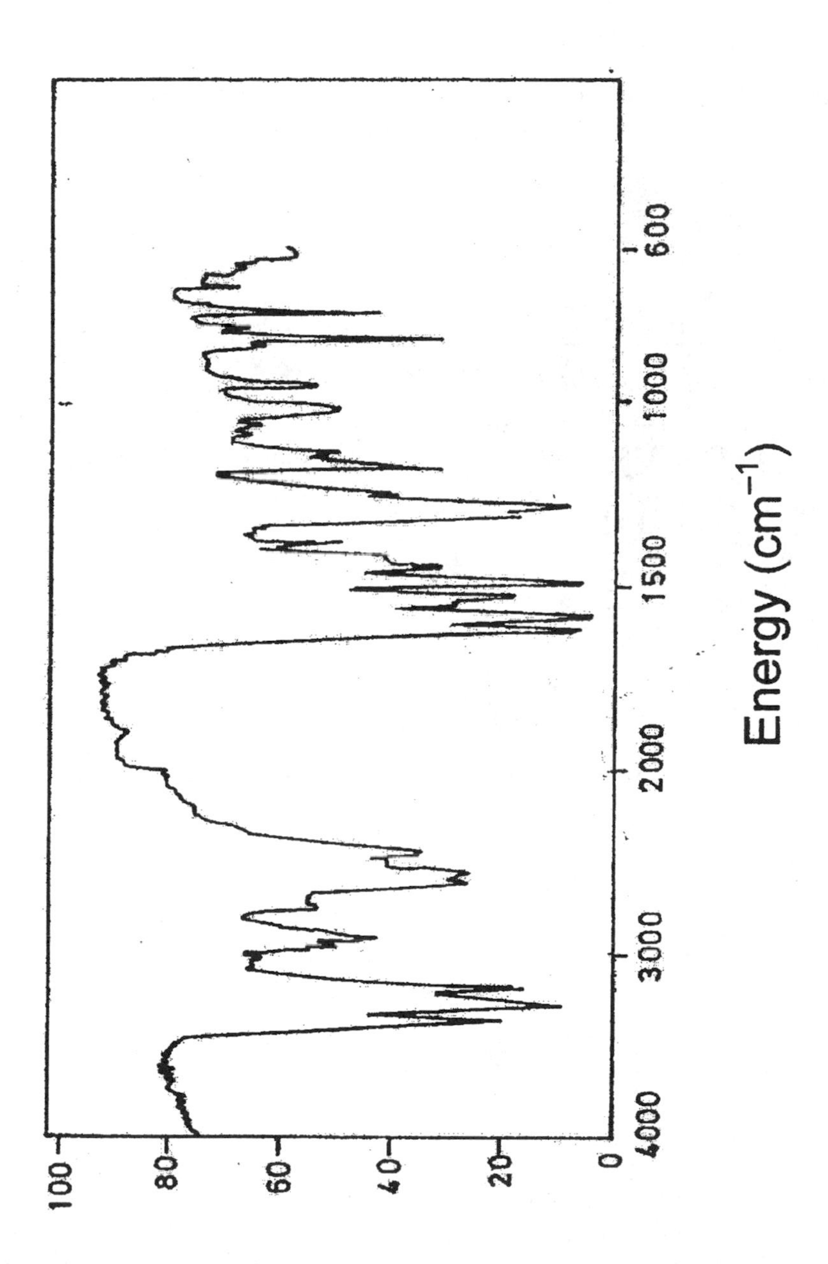

Table 2

Assignments for the Observed Infrared Frequencies of Procainamide Hydrochloride

Frequency (cm^{-1})	**Assignment**
3390, 3285, 3200	NH_2 stretch and NH stretch
2860	Aromatic CH stretch
1625	Amide C = O stretch
1585	Aromatic
1500	Aromatic C = C stretch
1485	C–O and C–N stretch
839	Para-substituted aromatic C–H stretch

3.6.2 Vibrational Spectroscopy

The infrared absorption spectrum of Procainamide hydrochloride was obtained in a KBr pellet, and recorded on a Pye Unicam SP spectrometer. The spectrum is presented in Figure 5, and assignments for the observed band frequencies and are presented in Table 2.

Clarke [9] reported principal infrared absorption peaks at 1600 cm^{-1}, 1512 cm^{-1}, 1639 cm^{-1}, 1297 cm^{-1}, 1545 cm^{-1}, 570 cm^{-1} when using the KBr pellet method for Procainamide HCl.

3.6.3 Nuclear Magnetic Resonance Spectrometry

3.6.3.1 ^{1}H-NMR Spectrum

The ^{1}H-NMR spectrum of Procainamide hydrochloride was recorded using a Varian XL200 200 MHz spectrometer. Tetramethylsilane (TMS) was used as the internal reference. The spectrum of the drug dissolved in DMSO-d_6 is shown in Figure 6, while the spectrum of the drug dissolved in DMSO-d_6 + D_2O is found in Figure 7. The chemical shifts and assignments for the observed resonance bands are presented in Table 3.

3.6.3.2 ^{13}C-NMR Spectrum

The ^{13}C-NMR spectrum of Procainamide hydrochloride (dissolved in DMSO-d_6) was recorded using a Varian XL200 200 MHz spectrometer. Tetramethylsilane (TMS) was used as the internal reference. The S2PUL pulse sequence ^{13}C-NMR spectrum is shown in Figure 8, while the APT and DEPT spectra are shown in Figures 9 and 10, respectively. A summary of chemical shifts and band assignments is found in Table 4.

3.7 Mass Spectrometry

The mass spectrum of Procainamide hydrochloride was obtained utilizing a Shimadzu PQ-5000 Mass Spectrometer, with the parent ion being collided with helium as the carrier gas. The mass spectrum thusly obtained is shown in Figure 11, where a base peak was observed at m/z = 86. Table 5 contains the proposed fragmentation pattern of the drug.

Clark [9] reported the following characteristic ions:

Principal peaks at m/z = 86, 99, 120, 30, 92, 87, 58, and 65

N-acetyl-Procainamide peaks at m/z = 86, 58, 99, 56, 162, 132, 149, and 205.

Figure 6. Full ^{1}H-NMR spectrum of Procainamide hydrochloride dissolved in DMSO-d_6.

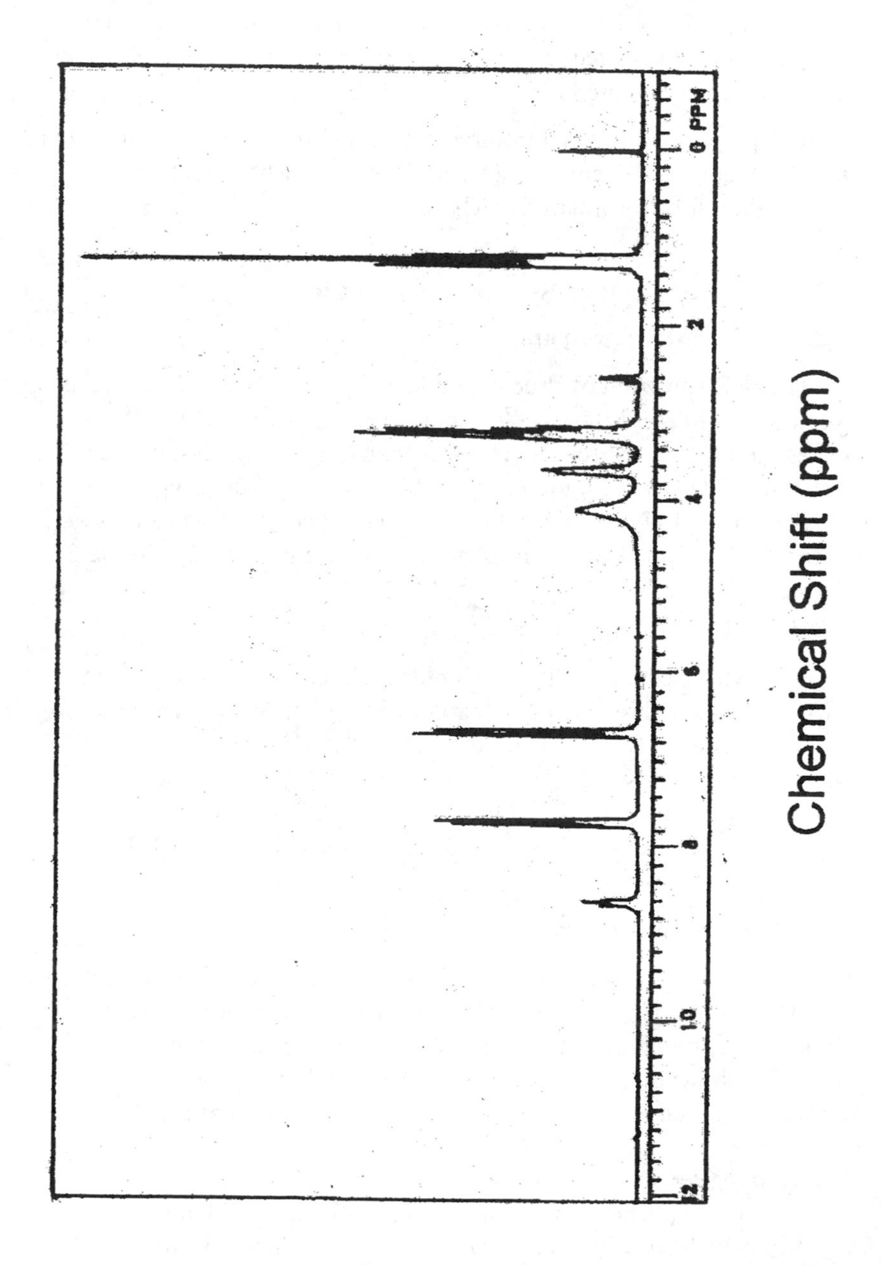

Figure 7. Full ^{1}H-NMR spectrum of Procainamide hydrochloride dissolved in DMSO-d_6 + D_2O.

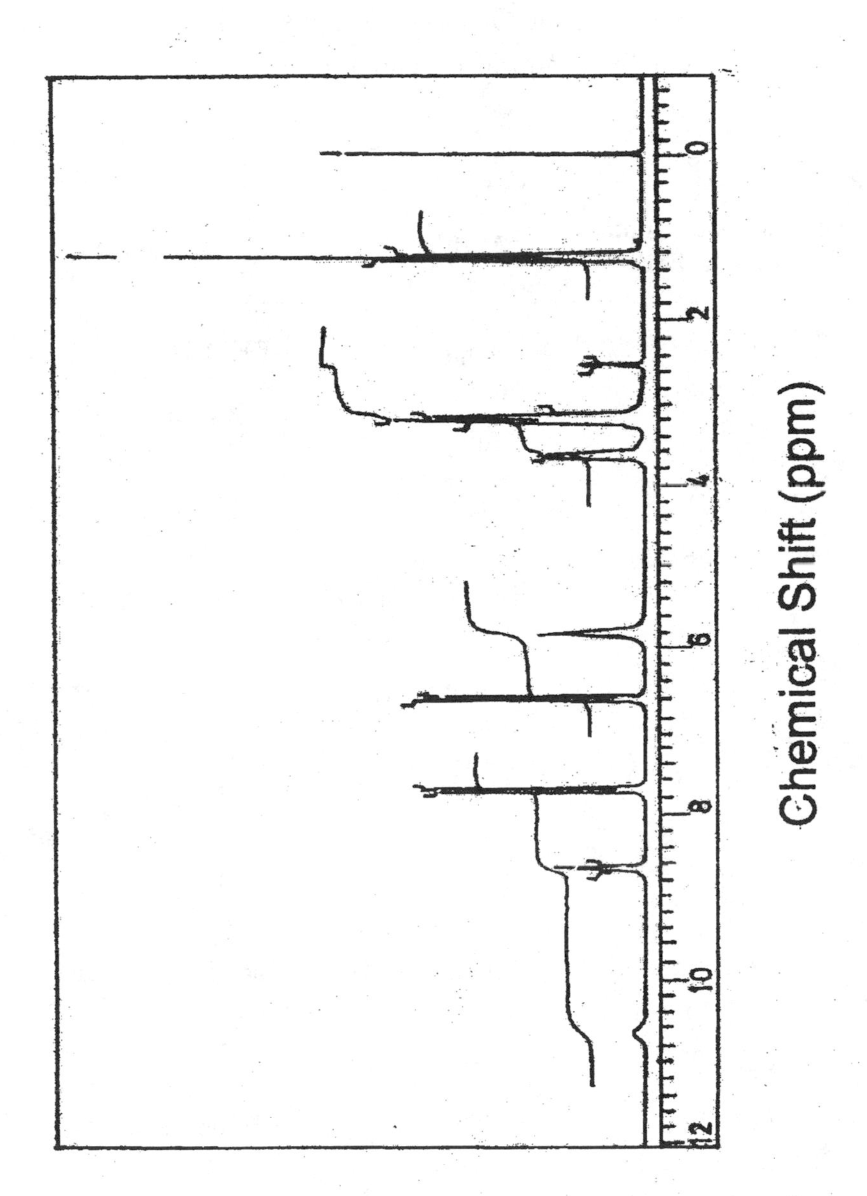

Table 3

Assignments for the Observed Resonance Bands in the ^{1}H-NMR Spectrum of Procainamide Hydrochloride

Chemical shift (ppm, relative to TMS)	**Multiplicity**	**Assignment**
1.3	Triplet (6H)	$-CH_2-\mathbf{CH_3}$
3.2	Multiplet (6H)	$-\mathbf{CH_2}-N-\mathbf{CH_2}-$ ‖ $\mathbf{CH_2}-$
3.6	Doublet (2H)	$-\mathbf{CH_2}-CH_2-$
5.8	Singlet (2H)	Exchangeable $N\mathbf{H_2}$
6.6	Doublet (2H)	$C_3-\mathbf{H}$
7.7	Doublet (2H)	$C_2-\mathbf{H}$
8.7	Multiplet (1H)	Partially exchanged $-C=N-$ $\mathbf{OH}$
10.6	NH (1H)	Exchangeable N**H**

Figure 8. ^{13}C-NMR spectrum of Procainamide hydrochloride dissolved in DMSO-d_6.

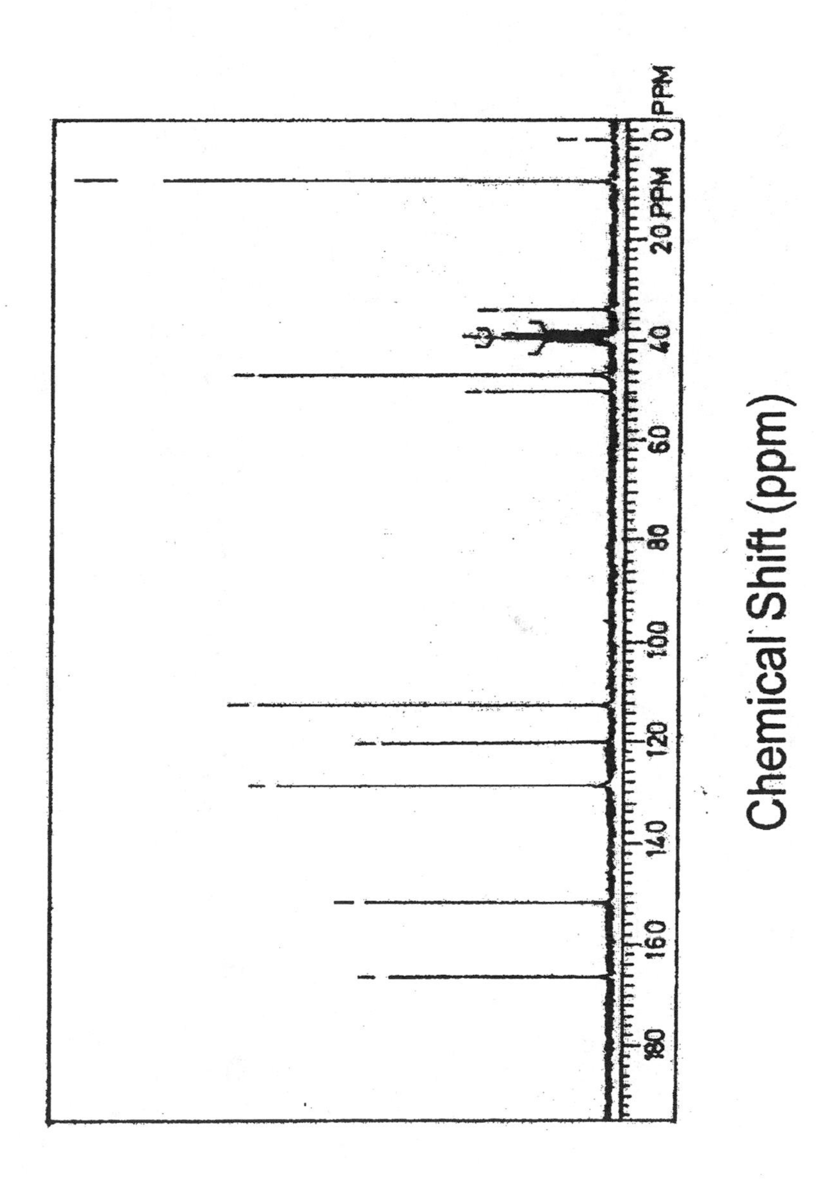

Figure 9. ^{13}C-NMR APT spectrum of Procainamide hydrochloride.

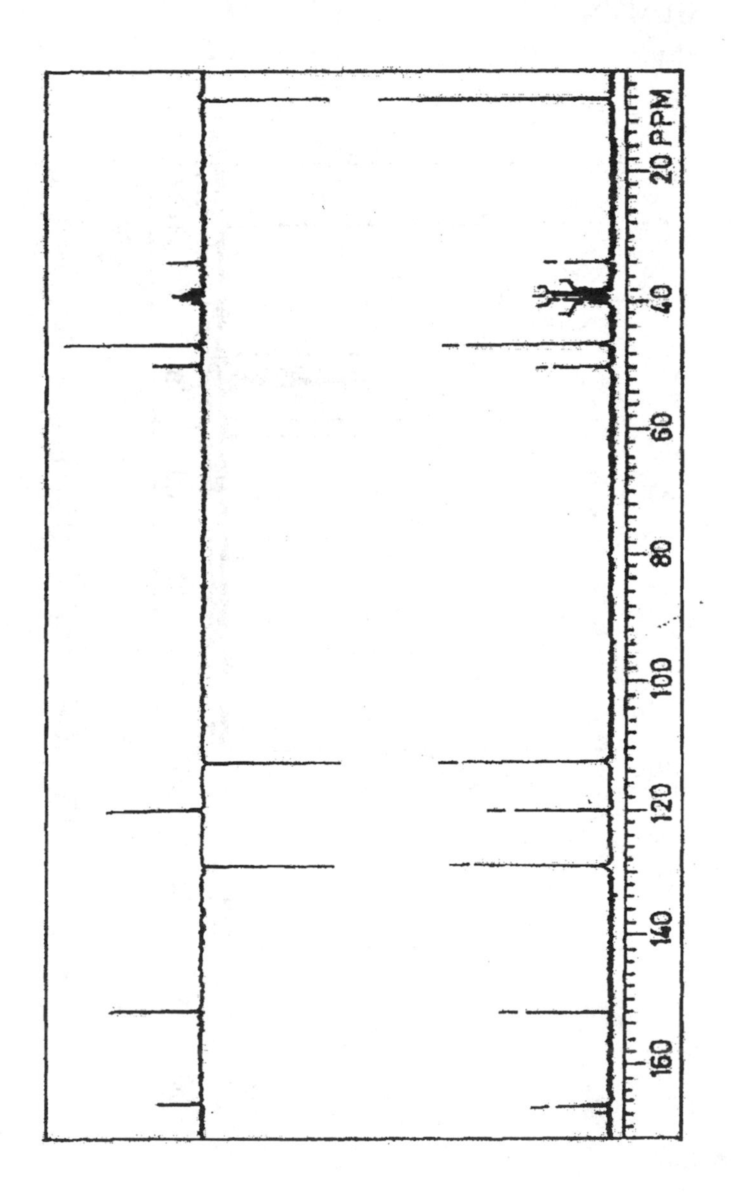

Figure 10. ^{13}C-NMR DEPT spectrum of Procainamide hydrochloride.

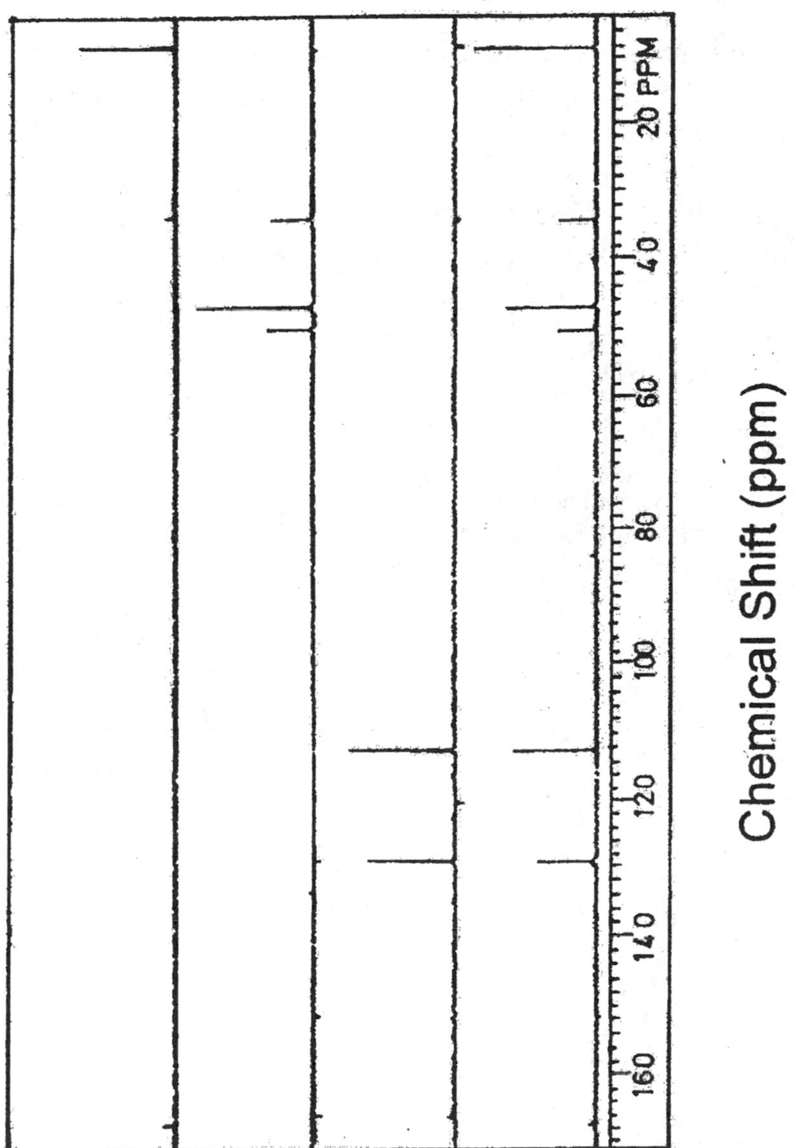

Table 4

Assignments for the Observed Resonance Bands in the ^{13}C-NMR Spectrum of Procainamide Hydrochloride

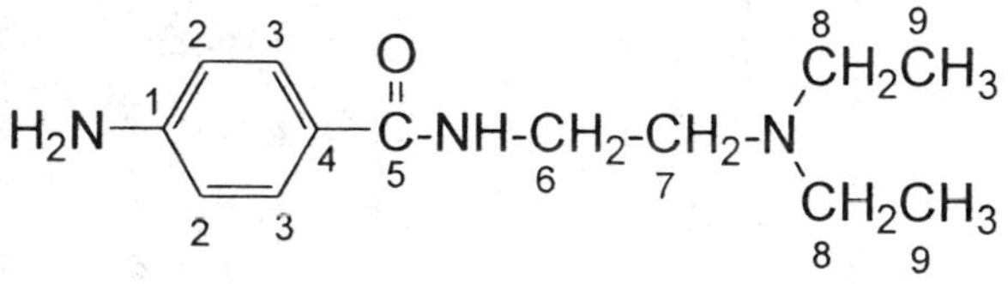

$H_2N-C_6H_4-C(=O)-NH-CH_2-CH_2-N(CH_2CH_3)_2$

Chemical shift (PPM, relative to TMS)	Carbon Number
8.4	9
33.9	8
46.8	7
50.1	6
112.5	3
120.2	2
128.8	1
151.9	4
166.5	5

Figure 11. Mass spectrum of Procainamide hydrochloride.

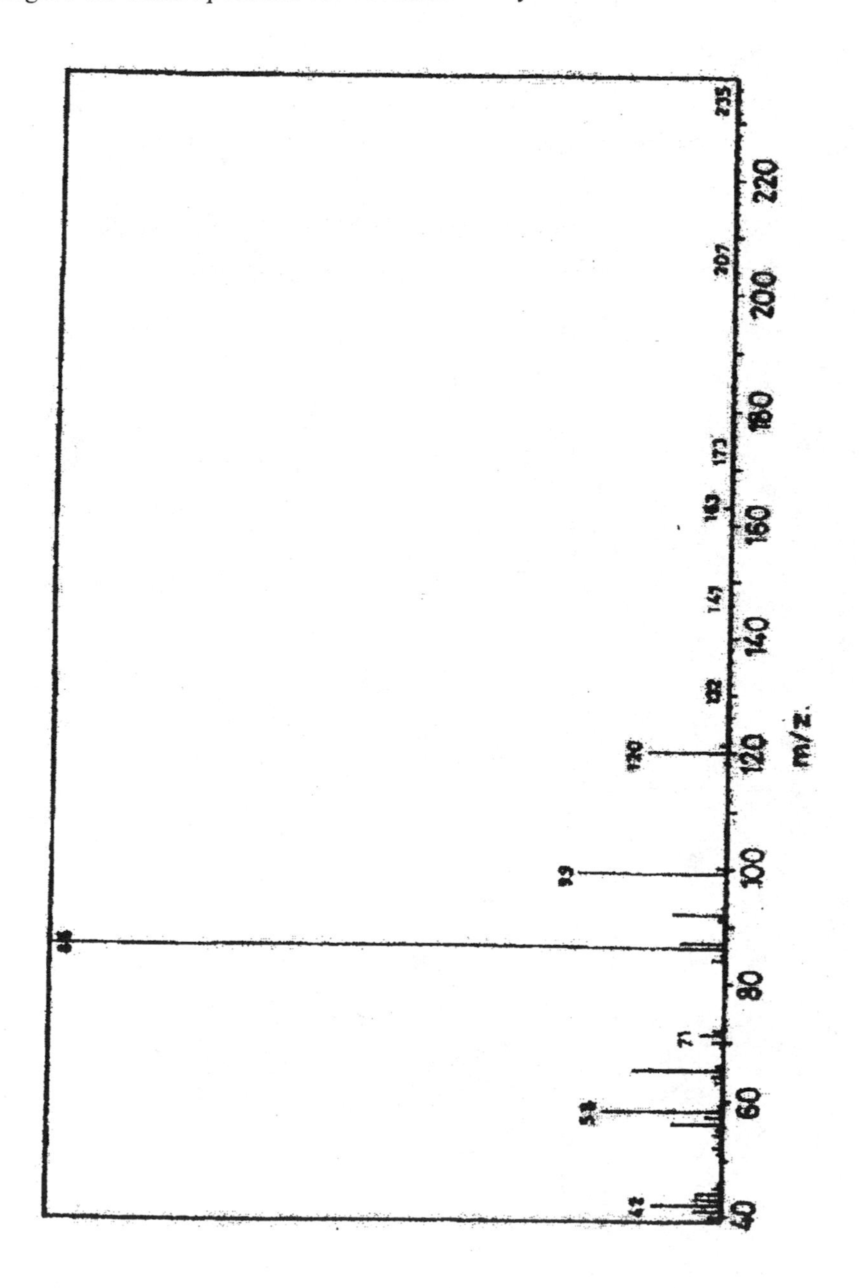

Table 5

Assignments for the Observed Fragmentation Bands in the Mass Spectrum of Procainamide Hydrochloride

m/z	Relative Intensity	Fragment
136	2%	$H_2N-C_6H_4-C(=\overset{+}{O}-H)-\dot{N}H$
120	12%	$H_2N-C_6H_4-C\equiv\overset{+}{O}$
100	2%	$\overset{\oplus}{C}H_2-CH_2-N(CH_2\text{-}CH_3)_2$
99	18%	$H_2C=CH-\overset{\bullet +}{N}(C_2H_5)_2$
92	8%	$H_2N-C_6H_4^{\oplus}$
86	100%	$H_2C=\overset{+}{N}(C_2H_5)_2$
65	14%	$C_5H_5^{+}$ (cyclopentadienyl cation)
58	15%	$H_2C=\overset{+}{N}(H)-CH_2\text{-}CH_3$
42	10%	$CH_3-C\equiv\overset{+}{N}-H$

4. Methods of Analysis

Kalashnikov and Mynka have reviewed the methods used for the determination of Procainamide and other aminocarboxylic acid with the intention to facilitate the rapid choice of an optimum procedure for a given analysis [64]. This review covers volumetric, optical, electrochemical, and chromatographic methods. Sterling *et al* reported a comparison between the fluorimetric, colorimetric, and gas liquid chromatographic methods [65].

4.1 Compendial Identification Tests

4.1.1 United States Pharmacopoeia 24 [2]

4.1.1.1 Infrared Absorbance. Equivalence of the infrared absorption spectrum of the analyte with that of the reference standard (KBr pellet method).

4.1.1.2 Thin-Layer Chromatography: Plates are coated with a 0.25 mm layer of chromatographic silica gel mixture, and the eluent is a 70:30:0.7 mixture of chloroform, methanol, and ammonium hydroxide. Methanol is used as a solvent for the test and for the standard solutions. Spots are visualized using ultraviolet light at 254 nm and at about 366 nm, followed by spraying with a 1 in 2000 solution of fluorescamine in acetone and viewing with ultraviolet light at 366 nm. The R_f value of the principal spot obtained from the test solution corresponds to that obtained from the standard solution.

4.1.2 British Pharmacopoeia 1993 [3a]

4.1.2.1 Melting Point: 166 to 170°C.

4.1.2.2 Light Absorption: The absorbance, in the range of 220 to 350 nm, of a 0.001% w/v solution in 0.1 M NaOH exhibits a maximum at 273 nm. The A(1%, 1 cm) at the maximum is 580 to 610.

4.1.2.3 Infrared Absorbance. Concordant with the spectrum of Procainamide hydrochloride EP-CRS.

4.1.2.4 Presence of Chloride: Dissolve 2.5 g of Procainamide hydrochloride in sufficient carbon dioxide-free water to produce 25 mL (*Solution S*). Dilute 1 mL of *Solution S* to 5 mL with water. Acidify with 2 M nitric acid, add 0.4 mL of silver nitrate solution, shake, and allow to stand. A curdy, white precipitate is produced.

4.1.2.5 Presence of Procainamide: Dilute 1 mL of *Solution S* to 2 mL with water. Acidify 1 mL of this solution with 2M HCl (or dissolve 0.1 g of the drug in 2 mL of 2M HCl) and add 0.2 mL of sodium nitrite solution. After 1 to 2 minutes, add the solution to 1 mL of 2-naphthol solution. An intense orange or red color and, usually, a precipitate of the same color are produced.

4.1.3 Additional Identification Tests

The British Pharmacopoeia 1988 [3] describes an additional identification test. Dissolve 1 g in 10 mL water, add 10 mL of 5 M NaOH, and extract with 10 mL of chloroform. Add 10 mL of toluene to the extract, dry over anhydrous sodium sulfate, and filter. Mix the filtrate with 5 mL of dry pyridine, add 1 mL of benzoyl chloride dropwise, heat on a water bath for 30 minutes, and pour into 100 mL of ether. Wash and allow to crystallize. The melting point of the crystals, after recrystallization from 45% aqueous ethanol, is about 186°C.

Procainamide was qualitatively differentiated from other drugs by a gel electrophoresis method, using a 1% agar-agar gel containing 0.02-0.1 M boric acid at 0.02 ma/cm^2 [66].

4.2 Titrimetric Analysis

4.2.1 Aqueous Titration

The 1988 British Pharmacopoeia describes a titrimetric method for the determination of Procainamide HCl injections [3]. To a volume containing 0.5 g of Procainamide HCl, add 45 mL of 6M HCl and boil for 1 minute. Cool and titrate with 0.1M sodium nitrite using 1 mL of ferrocyphen solution as indicator until a violet color is produced that is stable for not less than 3 minutes. Repeat the operation without the injection (blank titration). The difference between the titrations represents the amount of sodium nitrite required. Each milliliter of 0.1M $NaNO_2$ is equivalent to 0.02718 g of Procainamide HCl.

The United States Pharmacopoeia 20 [2a] described an aqueous titration procedure for the assay of Procainamide. Accurately weigh about 500 mg of Procainamide hydrochloride and transfer to a 250 mL conical flask. Add 50 mL of 6N hydrochloric acid, shake the flask for 3 to 5 minutes to attain complete solution, add 1 mL of Ferrocyphen indicator solution, and titrate promptly with 0.1 M sodium nitrite VS to a violet end point that is stable for not less than 3 minutes. Perform a blank titration on the reagents. The molarity is calculated using the formula:

$$[M] = S / \{ 172.50 (A - B) \}$$

where S is the quantity (in mg) of Procainamide hydrochloride, and A and B are the volumes (in mL) of 0.1 M sodium nitrite VS used in the respective Procainamide hydrochloride and blank titrations. Each milliliter of 0.1 M $NaNO_2$ is equivalent to 27.18 mg of $C_{13}H_{21}N_3O \cdot HCl$.

4.2.2 Non-Aqueous Titration

The American Chemical Society has specified a non-aqueous titration method [67]. Weigh approximately 150-170 mg of benzocaine into a 250 mL conical flask. Transfer 100 mL of acetic acid into the flask, and dissolve the benzocaine in the acid by placing the container in an ultrasonic bath at 20-25°C for 2-3 minutes. Add 10 mL of a mercuric acetate solution, and let the flask stand for 45 minutes. Add 2 drops of the crystal violet indicator, and titrate with 0.1N perchloric acid to a green end point with absence of blue shades. Accurately weigh about 135 mg of the Procainamide hydrochloride sample into the flask containing the neutralized benzocaine solution. Dissolve the Procainamide hydrochloride by placing the flask in an ultrasonic bath at 20-25°C for 2-3 minutes. Titrate with the standardized perchloric acid to a green end point with the absence of blue shades. The percent of Procainamide is calculated using:

$$\%\ \text{Proc.} = \{ (V) (N) (135.9) (100) \} / m$$

where V is the volume (in mL) of acetous perchloric acid needed to titrate the sample, N is the normality of acetous perchloric acid, m is the mass of sample (in mg) taken, and 135.9 is the equivalent weight of Procainamide hydrochloride.

Kadin reported that traces of acetic anhydride and acetaldehyde in the acetic acid titration solvent depress the non-aqueous titrimetry of primary aromatic amines [68]. Alternative methods include a modified non-aqueous titration and a nitrate titration with an internal ferrocyphen indicator. These alternative methods for the titration of Procainamide HCl were compared with those in USP XVIII for the non-aqueous titration of Procainamide. The nitrate titration method was found to be the most accurate.

Asahi *et al* titrated Procainamide hydrochloride after acetylation [69]. The drug was titrated with perchloric acid in acetic anhydride-anhydrous acetic acid (9:1 to 7:3) medium, in the presence of a glass electrode and a silver-silver chloride reference electrode with a ceramic junction containing saturated lithium chloride solution in anhydrous acetic acid.

4.2.3 Potentiometric Titration

Zommer-Urbanska and Urbanska determined Procainamide hydrochloride and some cardiac drugs using ion-selective electrode and a sliver electrode [70]. The drug was determined in tablets, dragees, or ampoules, using direct potentiometry with a chloride-selective electrode against a saturated calomel electrode. A linear calibration curve was obtained in the pCl range 1 to 4, or indirectly by potentiometric silver nitrate titration with a Ce-selective or silver indicator electrode. Statistical analysis showed no significant difference in results among the three methods.

The effect of Procainamide on the direct potentiometry of sodium, potassium, calcium, and chlorine were studied by Lewandowski *et al* [71]. Seronorm quality control serum (Nyegaard and Co., Oslo, Norway) was treated with the drug, adjusted to pH 7.4 with H_2SO_4 or Tris buffer, and the analysis was carried out using a Microlyte Analyzer (Kone Corp., Finland).

4.3 Electrochemical Analysis

4.3.1 Coulometry

Nikolic *et al* reported the preparation and the coulometric determination of various quaternary ammonium iodide derivatives of Procainamide, and of some other local anesthetics [99]. The quaternary iodide salts of

Procainamide were prepared by precipitation with methyl iodide. The quaternary iodide was then determined to a coulometric endpoint using a Radiometer titrator with a silver cathode and anode and a reference Hg_2SO_4 electrode.

4.3.2 Voltammetry

Bishop and Hussein studied the electrochemical analysis by anodic rotating disc electrode voltammetry in acidic or buffered media of Procainamide hydrochloride [100]. Electrode kinetic parameters for mass and charge transfer were determined, and the reaction mechanisms were elucidated.

Wheeler *et al* conducted electrochemical investigations on the immunologically reactive Procainamide metabolites [101]. The electrochemical characterization of the N-oxidized metabolites of Procainamide using cyclic voltammetry and liquid chromatography with electrochemical detection was described. Reaction characterization and binding behavior were described for each of the Procainamide metabolites following the *in vitro* incubation with several relevant biomolecules.

Katsu *et al* used an ion selective electrode method for the determination of Procainamide in blood serum [102]. The Procainamide-selective membrane electrode was prepared by using a PVC membrane impregnated with sodium tetrakis-[3,5-bis-(2-methoxyhexafluoro-2-propyl)phenyl] borate as the ion exchanger and 2-fluoro-2'-nitrodiphenyl ether as the membrane solvent. The membrane was attached to a PVC tube (4 mm o.d., 3 mm i.d.) that was filled with an internal solution of 1 mM Procainamide hydrochloride in 10 mM sodium chloride. A high input impedance voltmeter was used to measure the potential difference. All measurements were made at 25°C in 100 μL serum. A calibration graph was obtained up to 20 μg/mL of Procainamide, with the slope of the linear range (~2-20 μg/mL) being 57 mV per decade. The detection limit was 1.5 μg/mL.

4.4 Spectroscopic Analysis

4.4.1 Ultraviolet Spectrometry

Catania and King studied the physicochemical incompatibilities of Procainamide hydrochloride (Pronestyl) and other cardiovascular and psychotherapeutic agents, with sodium ethacrynate (Edecrin) [72]. The physical and chemical compatibility of sodium ethacrynate, combined with Procainamide hydrochloride in sodium chloride injection, was studied using pH measurements and spectrophotometric analyses. The therapeutic drug concentration of the admixture was inspected for particulate matter at one, four, and eight hours at room temperature, and the pH determination was made at the same time interval. Ultraviolet spectra were taken for the drug alone in sodium chloride injection and in combination in sodium chloride injection. The admixture spectrophotometric measurements were taken at one, four, and eight hours at room temperature. The ultraviolet spectra for sodium ethacrynate (Edecrin) - Procainamide hydrochloride (Pronestyl) admixture indicated that chemical reactions were involved.

Hamm developed an ultraviolet spectrophotometric method for the determination of Procainamide hydrochloride in capsules, tablets, and in injection dosage forms [73]. The sample was cleaned and extracted by solvent partitioning, and determined spectrophotometrically in 0.01 M sodium hydroxide at about 272 nm relative to a standard.

Terlingo *et al* reported an *in vitro* study on the compatibility of intravenous ranitidine and Procainamide [74]. Absorption spectra of the pure drug and pure ranitidine were first determined. Procainamide has one absorption peak at 290 nm, and therefore was chosen as the compound of analysis throughout the study. The absorbance of ranitidine did not interfere at this wavelength. Procainamide, in a concentration of 2 gm/500 mL (4 μg/mL) was analyzed alone and in combination with ranitidine 50 mg/100 mL (500 μg/mL). An incompatibility would be observed by changes in absorption spectra with the combination of both drugs versus Procainamide alone. Procainamide and ranitidine, in the concentrations studied, were compatible over a 24-hour period based on data from ultraviolet spectrophotometry. No haze, precipitate, color change, or evolution of gas was observed.

4.4.2 Photometry

A small laboratory analyzer was used by Silver for the therapeutic monitoring of Procainamide and some other drugs [98]. The analyzer consists of a fluid delivery module and an optics-mixer module with a standard filter photometer, a tungsten-halogen lamp, and a photodiode detector. The photometer uses polychromatic light. The disposable optical cells contain dry reagents for the specific drug test, and require water as diluent. The mean run-to-run coefficient of variation for Procainamide and the other drugs are 10%.

4.4.3 Colorimetry

Plavsic determined Procainamide in the charcoal extractions of acidic, neutral, and basic drugs from human serum by spectrophotometric and fluorimetric methods [76].

Klitgaard described a colorimetric method for the routine and simultaneous determination of Procainamide and N-acetyl-Procainamide in serum [77]. The method involves extraction with dichloromethane and re-extraction with l N HCl. Procainamide was determined in the hydrochloric acid extract by a Bratton-Marshall reaction, and the N-acetyl-Procainamide concentration was obtained as the difference between the Procainamide concentrations measured after and before heating the acidic extract in a boiling water bath for 10 minutes. Concentrations as low as 0.2 μg/mL could be determined.

Tan *et al* reported a colorimetric assay method for Procainamide hydrochloride and some dosage forms using 4-dimethylamino-cinnamaldehyde as the colorimetric reagent [78].

A simple colorimetric reaction was described by Whitaker and Hoyt that was used for the determination of Procainamide in injectable preparations by complexing the hydrochloride with Cu(II) [79]. A few drops of 2M Cu(II) solution were added to a solution of Procainamide hydrochloride (1 mL) at pH 4 to 4.5 (acetate buffer), and the absorbance was measured at 380 nm.

Sastry *et al* described a colorimetric method for the assay of Procainamide in tablets and injection solution [80]. The drug reacts with 2-aminophenol and its N-substituted derivatives in the presence of oxidizing agents (e.g. KIO_3 solution) to give red compounds which can be determined colorimetrically at 520 nm.

Filipeva *et al* reported a colorimetric method for the determination of Procainamide in which a sample of the drug solution or tablet dissolved in water was mixed with 0.01M sodium 1,2-naphthoquinone-4-sulfonate, and the absorbance measured at 485 nm [81]. The concentration of the drug was found by comparison with the absorbance of a standard solution. The same authors also reported another colorimetric method for the determination of Procainamide in its dosage forms, based on drug reaction with sodium 1,2-naphthoquinone-4- sulfonate. The absorbance was measured at 480 nm from an ethanol solution [82].

Rao *et al* determined the drug spectrophotometrically at 765 nm after treatment with Folin-Ciocalteu reagent in the presence sodium carbonate [83].

Sastry *et al* reported a procedure for the spectrophotometric determination of Procainamide HCl after extraction of the drug with water [84]. Aliquots of these solutions are treated with 0.8% $FeCl_3$ solution in 0.01M-HCl and aqueous 0.2% 3-methyl benzothiazolin-2-one hydrazone hydrochloride. The resulting color was measured a reagent blank at 580 nm.

Mohamed *et al* used 7,7,8,8-tetracyano-quinodimethane for the spectrophotometric determination of Procainamide [85]. Aqueous sample solutions were treated with 0.12% of tetracyanoquinodimethane solution in acetonitrile and 0.15 M sodium carbonate. The solution was further diluted with water and set aside for two minutes before the absorbance of the solution was measured at 473 nm against a reagent blank.

Sitar *et al* reported a modified colorimetric method for the determination of Procainamide in plasma [86]. The method involves adding plasma to sodium chloride and sodium hydroxide, and then shaking the contents with dichloromethane. After centrifugation, a specified volume of organic layer is removed and evaporated under a stream of nitrogen. After cooling, the contents are mixed with HCl, ammonium sulfamate, and Marshall's Reagent. The solution is left to stand at room temperature and its absorbance measured at 550 nm against a plasma blank.

Gurkan reported the determination of Procainamide HCl by reaction with sodium 1,2-naphthoquinone-4-sulfonate in an aqueous medium at pH 3.5 [87]. The resulting chromophore yielded an absorption maximum at 485 nm. The absorption was linear over a range of 50-250 μg/mL and stable for 24 hours.

Minka *et al* used an absorption method for the determination of Procainamide [88]. The drug was derivatized by diazotization, coupling with 3-(1,2 dicarboxyethyl) rhodamine solution. Absorptiometry (blue filter) of the red solution obtained by addition of sodium hydroxide solution. A calibration graph was used for evaluation, with Beer's law being obeyed over the range of 1 to 10 μg/mL Procainamide.

Sterling *et al* conducted a comparative study on the determination of Procainamide using fluorimetry, colorimetry, and GLC methods [65]. Finally, a colorimetric method for the determination of Procainamide was also reported by Peters [89].

4.4.4 Infrared Spectrometry

Mynka *et al* reported an infrared spectroscopic method for the determination of Procainamide and other drugs of the amide class [75].

4.4.5 Fluorimetry

The literature reports several fluorimetric methods for the determination of Procainamide in dosage forms, serum, or plasma [76,90-95]. Plavsic reported the determination of the drug, in the charcoal extractions of acidic, neutral, and basic drugs from human serum by a fluorimetric method [76]. Tan and Beiser reported a rapid fluorimetric method for the determination of Procainamide hydrochloride in dosage forms [90]. Tan *et al* also determined the drug by a spectrophotofluorimetric method [91]. Pape also reported the determination of the drug in serum by a spectrofluorimetric method [92].

Sterling and Haney described a spectrophotofluorimetric method for the analysis of the drug in the presence of primary aliphatic amines based on reaction with fluorescamine [93]. Frislid *et al* used a fluorimetric and a gas chromatographic method for determination of the drug [94]. Ambler and Masarei reported analysis of the drug by a fluorimetric method [95].

As mentioned earlier, Sterling *et al* conducted a comparative study on spectrofluorimetric, colorimetric, and gas liquid chromatographic methods for the analysis of Procainamide in plasma [65].

4.4.6 Phosphorescence

The determination of Procainamide by after derivatization and room temperature phosphorescence has been reported by Long *et al* [96]. In this work, the drug was derivatized with fluorescamine, without prior separation of the primary amino active ingredients, in pharmaceutical preparations. The optimum pH, buffer concentration, and phosphorescence characteristics were discussed.

4.4.7 Flow Injection Spectrophotometry

Sultan and Suliman reported the use of a flow injection spectrophotometric method for the determination of Procainamide HCl in proprietary drugs [97]. The method is based on the oxidation of Procainamide hydrochloride with Ce(IV) in sulfuric acid, and subsequent monitoring of the absorbance at 480 nm [97]. The Factorial Design program was used to examine the significance of sulfuric acid concentration, Ce(IV) concentration, sample loop size, flow rate, and coil length on the sensitivity and the conditions were optimized by the Super Modified Simplex program. Calibration graphs were linear over the range of 100 to 600 ppm.

4.5 Chromatographic Methods of Analysis

4.5.1 Paper Chromatography

Clarke [9] reported three paper chromatography systems.

4.5.1.1 System 1

This system is based on the use of Whatman No. 1 paper, which is buffered by dipping in a 5% solution of sodium dihydrogen citrate, blotted, and dried at 25°C for 1 hour. The sample consists of 2.5 μL of a 1% analyte solution in 2 N acetic acid if possible, but otherwise in 2 N hydrochloric acid, 2 N sodium hydroxide, or ethanol. The developing solvent consists of 4.8 g of citric acid in a mixture of 130 mL of water and 870 mL of *n*-butanol and was developed ascending in a 8 × 11 × 15½ inch tank, 4 sheets being run at a time. A 5-hour time period was required to complete the analyses. Visualization was effected using ultraviolet light (blue fluorescence), iodoplatinate spray (strong reaction), or bromcresol green spray (strong reaction). The R_f value was reported to be 0.19 [105].

4.5.1.2 System 2

This system is based on the use of Whatman No. 1 paper impregnated by dipping in a 10% solution of tributyrin in acetone and which is dried in air. The sample consists of a 5 µL aliquot of 1 to 5% analyte solution in ethanol or chloroform. The developing solvent consists of acetate buffer (pH 4.58). The beaker containing the solvent is equilibrated in a thermostatically controlled oven at 95°C for about 15 minutes, and was developed ascending. The paper is folded into a cylinder and clipped, the cylinder inserted in the beaker containing the solvent (which is not removed from the oven). A plate-glass disc thickly smeared with silicone grease may serve as a cover. The time for a typical run is 15 to 20 minutes. Visualization is effected under ultraviolet light (blue fluorescence) or using Iodoplatinate spray (positive reaction). The R_f value was reported to be 0.92 [106].

4.5.1.3 System 3

In this system the paper and sample preparation are the same as described above for System 2, but with the solvent consisting of phosphate buffer (pH 7.4). The beaker containing the solvent is equilibrated in a thermostatically controlled oven at 86°C for about 15 minutes. The development and spot location are as for system 2. The R_f value was reported to be 0.90 [107,108].

4.5.2 Thin-Layer Chromatography

Clarke reported a TLC system [9a]. Plates are coated with a slurry consisting of 30 gm of silica gel G in 60 mL of water, yielding a layer 0.25 mm thick that is dried at 110°C for 1 hour. The sample consists of 1.0 µL of a 1% solution in 2N acetic acid. The mobile phase is 1.5:100 strong ammonia solution / methanol, which should be changed after two runs. The solvent is allowed to stand in the tank for 1 hour, and is developed ascending (tank 21 × 21 × 10 cm), the end of the tank being covered with filter paper to assist evaporation. The time of a run is 30 minutes. The developing agent is acidified iodoplatinate spray, for which a positive reaction is obtained. The R_f value was reported to be 0.45 [109].

Reidenberg *et al* determined the concentrations of Procainamide and its N-acetyl-Procainamide metabolite in the plasma and in urine of patients of known acetylator phenotype (dapsone phenotyping) who had been taking Procainamide for more than three days [46]. These workers used .both the thin layer chromatographic and the densitometry methods

A thin layer chromatography method for the estimation of Procainamide and its major metabolite in plasma, was described by Gupta *et al* [110]. The method involves a plasma extraction at alkaline pH with dichloromethane to isolate Procainamide and its metabolite, N-acetyl-Procainamide. *p*-Amino-N-(2-dipropylaminoethyl) benzamide and its N-acetyl derivative were added as internal standards to the plasma prior to extraction. A solvent system was selected which allowed excellent separation of the four compounds from each other, from plasma constituents, and from other drugs commonly prescribed with Procainamide. It was demonstrated that densitometric scanning in the fluorescent mode was a precise and specific technique to quantitate Procainamide and its N-acetyl metabolite.

Kark *et al* reported the use of a thin layer chromatographic method for the determination of Procainamide and its N-acetyl metabolite in human serum and urine at single dose levels [111]. Serum and urine levels of Procainamide hydrochloride and N-Acetyl-Procainamide hydrochloride were determined *in vitro* and in five subjects using thin layer chromatography assay with an ultraviolet detection.

Non-linear calibration in quantitative high performance thin layer chromatography was used by Sistovaris for the analysis of Procainamide in urine [112]. The non-linear parameter procedure of Kufner and Schlegel, and a log-log linear regression procedure, were compared with the use of Procainamide and its N-acetyl metabolite in urine as a model sample [113]. The agreement between the calibration and values for standards in the former method was in the range of 1 to 3% of the measured value, and in the latter method the range was 2 to 10%.

Sistovaris *et al* reported the determination of Procainamide and its N-acetyl metabolite in human serum and in human urine (at single-dose levels) by thin layer chromatography [114]. The drug and its metabolite were extracted from serum at pH 10.9 into ethyl acetate, then the residue from evaporation of the extract was dissolved in ethyl acetate. This was applied to silica gel F_{254} HPTLC plates for thin layer chromatography, eluted with 9:1 dioxane / concentrated aqueous ammonia, and scanned at 275 nm. This method was used in a pharmacokinetic study of Procainamide in man.

High performance thin layer chromatography (HPTLC) was used by Malikin *et al* for the monitoring of Procainamide and other therapeutic drugs [115]. For the Procainamide assay, serum was extracted at pH 10 with 1:1 dichloromethane /heptane containing 10 ppm of octanol. One quarter of the organic phase was applied to one well of a HPTLC contact Spotter and evaporated. Standard samples were processed similarly, so that 15 solutions could be processed simultaneously. The plates were developed for 30 minutes with 7:6:1:1 benzene/ ethyl acetate/ methanol/ triethylamine, and then dried. The reflectance was scanned at 265 nm.

In order to improve the applicability of reversed phase thin layer chromatography to the estimation of drug lipophilicity, the pH dependence of thermodynamically true RM value was investigated by Dross *et al* [116]. A 0.5 μL volume of an ethanolic solution of Procainamide was applied to an RP_{18} $F_{254}S$ layer. The front markers were KBr, KI, and $NaNO_3$, and the mobile phases were methanol-buffer mixtures. The plates were evaluated using densitometry.

Clarke [9] described three systems for the TLC analysis of the drug. All three systems use the same separation place, which is silica gel G, spread in a layer 250 μm thick, and which has been dipped in or sprayed with 0.1 M KOH in methanol and dried. In one method, the mobile phase is 100: 1.5 methanol / strong ammonia solution. The reference compounds are diazepam (R_f 75), chlorprothixene (R_f 56), codeine (R_f 33), and atropine (R_f 18). The R_f of Procainamide in this method is 49 [117]. In the second method, the mobile phase is 75:15:10 cyclohexane / toluene / diethylamine, and the reference compounds are dipipanone (R_f 66), Pethidine (R_f 37), desipramine (R_f 20), and codeine R_f 06. The R_f of Procainamide in the second method is 01. [117]. In the third method, the mobile phase is 90:10 chloroform / methanol, and the reference compounds are meclozine (R_f 79), caffeine (R_f 58), dipipanone (R_f 33), and despiramine (R_f 11). The R_f of Procainamide 5, after being visualized with acidified iodoplatinate solution [117].

Wesley-Hadzija and Mattocks reported the use of a quantitative TLC method for the determination of Procainamide and its major metabolite in plasma [118]. Lee *et al* described a simultaneous determination of Procainamide and other anti-arrhythmic drugs using a HPTLC method [119].

4.5.3 Gas Chromatography

A gas liquid chromatography (GLC) method for the determination of Procainamide in biological fluids was described by Simons and Levy [120]. By using the dipropyl analog of Procainamide as an internal standard, both compounds could be chromatographed directly, yielding linear calibration curves and a sensitivity that allows quantitative determination of concentrations as low as 0.1 μg/mL.

Simons *et al* developed a specific gas chromatographic method for the determination of Procainamide in biological fluids [121]. The method allowed the study of the pharmacokinetics of the drug in several normal subjects. Carter *et al* used a GLC method for the analysis of Procainamide HCl in rat feed [122].

A GLC method, using a nitrogen-phosphorous selective detector, was reported by Kessler *et al* for the simultaneous quantitative determination of Procainamide, and its N-acetyl metabolite, in serum [123].

Yamaji *et al* described a method for the simultaneous determination of Procainamide and its N-acetyl metabolite in serum using gas chromatography with nitrogen-selective detection [124]. Serum samples were placed in glass culture tubes (7.5 cm x 12 mm) and mixed with 2M sodium hydroxide. The alkaline solution was applied to an Extrelut column, and after 15 minutes Procainamide and N-acetyl-Procainamide were eluted with dichloromethane. A methanolic solution of propranolol was added to the eluent as an internal standard. The solution was evaporated to dryness and the residue dissolved in methanol. A portion of the solution was injected onto a glass column (2.4 m x 2 mm) packed with 3% OV-17 on Chromosorb W (80 to 100 mesh). The carrier gas was nitrogen at 30 mL/minute, and temperature programming from 225°C to 260°C.

Sterling *et al* developed procedures for the analysis of Procainamide in plasma, using spectrophotofluorimetry, colorimetry, and gas liquid chromatography after the reaction of Procainamide with fluorescamine [65]. The accuracy and precision of results obtained using these methods were compared to those obtained using other colorimetric and gas liquid chromatography methods. The utility of the three procedures in the routine determination of Procainamide in plasma was discussed.

The micro determination of Procainamide HCl in human serum with an electron capture gas chromatographic method was reported by Ludden *et al* [125]. The method was studied in spiked samples of human blood and *in vivo*.

Clarke reported two GC systems for Procainamide [9]. In the first system, the column is 2.5% SE-30 on 80-100 mesh Chromosorb G (acid-washed and dimethyldichlorosilane-treated), 2m×4 mm i.d. glass column. It is essential that the support is fully deactivated. The column temperature is between 100°C and 300°C, and the carrier gas is nitrogen at 45 mL/min. Suitable reference compounds are *n*-alkanes with even number of carbon atoms. The retention index of Procainamide is 2248, and that for N-acetyl-Procainamide is 2698 [126]. In the second system, the column is 3% Poly A 103 on 80-100 mesh Chromosorb W HP, 1 m × 4 mm i.d. glass column. The column temperature is 200°C, and the carrier gas is nitrogen at 60 mL/min. Suitable reference compounds are *n*-alkanes with an even number of carbon atoms. The retention index of Procainamide is 2965 (80) [127].

Clarke has described another GC system [9a]. Here the column is 2.5% SE-30 on 80-100 mesh Chromosorb WA WHMDS, 5 feet × 4 mm i.d. glass column. The column temperature is 225°C, and the carrier gas is nitrogen at a flow rate of 50 mL/min. Detection is by flame ionization, using hydrogen at 50 mL per minute and an air flow of 300 mL per minute. The retention time is 0.63, relative to codeine.

Zhuravleva *et al* determined Procainamide in pharmaceutical formulation by capillary gas chromatography, using a column (27 m × 0.3 mm i.d.) coated with OV-101/KF or a column (20 m × 0.3 mm i.d.) coated with Triton X-305 operated. The method called for temperature programming from 120-200°C at 8°C/min, with helium as the carrier gas and a flame ionization detector [128].

Frislid *et al* reported the use of fluorimetric and gas liquid chromatographic methods for the analysis of Procainamide [94].

4.5.4 High Performance Liquid Chromatography

USP 24 describes a HPLC method for the assay of Procainamide hydrochloride [2]. The mobile phase used consists of a 140:60:1 mixture of water / methanol / triethylamine, adjusted with phosphoric acid to a pH of 7.5, that is filtered and degassed. The chromatography system used is equipped with a 280 nm detector and a 3.9 mm × 39 cm column that contains 10 μm packing of octadecyl silane chemically bonded to porous silica or ceramic microparticles, 3 to 10 μm in diameter. The mobile phase flow rate is about 1 mL/min. The standard solution is prepared by dissolving an accurately weighed quantity of USP Procainamide hydrochloride RS quantitatively in the mobile phase to obtain a concentration of 0.5 mg/mL, and diluting with mobile phase to obtain 0.05 mg/mL. The assay preparation is made by transferring about 50 mg of Procainamide hydrochloride to a 100 mL volumetric flask, dissolving in and diluting with the mobile phase to volume. The solution is mixed and further diluted in an another volumetric flask with the mobile phase to volume and mixed. In the method procedure, one separately injects equal volumes (about 20 μL) of the standard preparation and the assay preparation onto the chromatograph, records the chromatograms, and measures the responses for the major peaks. The quantity, in mg, of $C_{13}H_{21}N_3O{\cdot}HCl$ in the sample of Procainamide hydrochloride taken is calculated using:

$$mg(proc) = \{ 1000\ C \} \{ r_u / r_s \}$$

where C is the concentration (mg/mL) of USP Procainamide hydrochloride RS in the standard preparation, and r_u and r_s are the peak responses obtained from the assay preparation and the standard preparation, respectively.

A novel polymeric reversed-phase sorbant for solid-phase extraction was reported by Bouvier *et al* [129]. Porcine serum was spiked with Procainamide at a concentration of 10 μg/mL. Three replicate extractions were performed for each sample on C_{18} and HLB SPE cartridges with 1 mL methanol. A 20-position vacuum manifold was used with a vacuum pump. Internal standard, sulfanilamide (10-1000 μL), was added to the eluent, and after vortexing the mixture was analyzed by HPLC. Three resins (30 mg) were used to determine the effect of divinylbenzene-*N*-pyrrolidone incorporation on the extraction performance: (1) poly(divinylbenzene), (2) a copolymer of divinylbenzene and vinylpyrrolidone with a mean particle size of 80 μm, a surface area of 786 m^2/g and a mean pore diameter of 180 angstrom, and (3) a novel hydrophilic-lipophilic balanced copolymer of N-vinylpyrrolidone and divinylbenzene. PBS (1 mL) was loaded spiked with one or more model compounds. The cartridges were washed with 1 mL of water and the analytes were removed from the sorbant with 1 mL aliquots of methanol. The cartridges were found to require a conditioning step with a water-miscible organic solvent before sample application. After the sorbant was wetted, water or aqueous buffer could displace the already filled pores.

Lessard *et al* described an improved high performance liquid chromatographic assay method for the determination of Procainamide and its N-acetylated metabolite in plasma, and the application of thismethod to a single dose pharmacokinetic study [130]. The concentrations were observed up to 32 hours after a single oral dose administration of the drug to human subjects. Following liquid-liquid extraction of plasma samples, the drug and its N-acetyl derivative, along with the internal standard N-propionyl-Procainamide, were separated on a reversed-phase C_8 column with retention times of 4, 6.7, and 13.2 minutes, respectively. The UV detection limit at 280 nm of Procainamide and its metabolite was 2 ng/mL (signal-to-noise ratio 3:1), and the quantitation limit was 4 ng/mL (signal-to-noise ratio 5:1). Intra- and interday coefficients of variation were less than 8% within the range of 20-500 ng/mL.

Anti-arrhythmics, according to a Chrompack application note, can be determined by HPLC on a Inertsil 5 ODS-2 ChromSep stainless steel column (25 cm × 4.6 mm i.d.) operated at 35°C with 20 mM phosphate buffer of pH 7.5/methanol (3:2) as mobile phase (2 mL/min) and ultraviolet detection at 254 nm. A chromatogram (run time 6 minutes) was presented, with peak identifications of Procainamide and N-acetyl Procainamide [131].

A comparison of the performance of the conventional C_{18} phases with others of alternative functionality for the analysis of Procainamide and other basic compounds by reversed-phase HPLC was reported by McCalley [132]. The performance of ten columns (all 25 cm × 4.6 mm i.d., with 5 μm particles), including the conventional C_{18} phases, shorter alkyl chains, cyano phases, phases with embedded polar groups, and a phase with increased stability at higher pH values were compared. The solutes, with a range of pKa values, were Procainamide and other basic drug. The mobile phases were mixtures of acetonitrile or methanol with water or phosphate buffers of varying molarity and of pH 3 or 7. Samples (0.2 μg) of single compounds in 2 μL of the mobile phase were injected, the columns were at 30°C, and solutes were detected at 215 or 254 nm.

Several other HPLC methods have been reported for use in the analysis of Procainamide hydrochloride and its metabolites [55,92,104,133-180]. The drug and its N-acetyl metabolite were determined in serum [92,133-139], in plasma [140-145], and in pharmaceutical dosage forms [146]. The conditions and systems of some of the other methods [55,104,147-180] are presented in Table 6.

4.5.5 Micellar Liquid Chromatography

DeLuccia *et al* used a direct serum injection, with micellar liquid chromatography, for therapeutic drug monitoring [189]. Detection limits were obtained for Procainamide and other eight commonly administered drugs well below the normal therapeutic range when these were determined by liquid chromatography in untreated serum. 0.02 - 0.10 M sodium dodecyl sulfate was used as the mobile phase, and the column was Supelcosil LC-CN. The results showed good agreement with those obtained using EMIT Kits.

4.5.6 High Performance Liquid Chromatography - Electrospray Ionization Mass Spectrometry

Needham and Brown reported a high performance liquid chromatography-electrospray ionization mass spectrometric (HPLC-ESIMS) method for the analysis of Procainamide and other basic pharmaceuticals [190]. Cyanopropyl and pentafluorophenylpropyl modified silica columns gave good retention and good peak shape for the HPLC-ESIMS analysis of several classes of basic drugs. The phases were found to enhance the ESIMS signal by providing good retention of basic drugs when operated with a mobile phase containing 90% acetonitrile. In order to achieve good retention of basic drugs, only a mobile phase containing less than 40% CH_3CN can be used with C_{18} columns.

Table 6

High Performance Liquid Chromatography Systems Used in the Analysis of Procainamide Hydrochloride

Column	Mobile phase and flow rate	Detection	Remark	Ref.
10 cm × 4 mm Novapak reversed-phase (5 μm) equipped with a 2 cm guard column for RP column (37 to 53 μm).	Water-methanol-acetic acid-triethylamine (74:25:1:0.03) (1 mL/min.).	280 nm	Longer plasma half-life for Procainamide utilizing HPLC assay in plasma.	[55]
25 cm × 4.6 mm of ultrasphere ODS (5 μm) with a guard column (3 cm × 4 mm) of μBondapak C_{18} – Corasil (37-50 μm)	0.05M KH_2PO_4 (pH 2.5) – acetonitrile (1 mL/min.)	220 to 270 nm	A micro dialysis method for the determination of plasma protein binding of Procainamide and other drugs.	[104]
12.5 cm × 4.9 mm of Sherisorb S5W	Methanolic 10 mM NH_4ClO_4 and methanolic 0.1 M-sodium hydroxide (pH 6.7)	254 nm followed by electro-chemical detection.	Analysis of the drug among other many basic drugs.	[147]

Table 6 (continued)

High Performance Liquid Chromatography Systems Used in the Analysis of Procainamide Hydrochloride

15 cm × 3.9 mm of Novapak C_{18} µm)	Acetic acid-water-acetonitrile (2:50:5) containing 3.5% of sodium acetate (1 mL/min.).	254 nm	Analysis of the drug and its N-acetyl metabolite in human serum.	[148]
10 cm × 5 mm of Novapak cyano HP.	5mM acetate buffer (pH 6) containing 0.05% triethylamine and 10% acetonitrile (1.5 mL/min.)	280 nm	Analysis of the drug and its N-acetyl metabolite in serum or plasma.	[149]
15 cm × 4.6 mm of spherisorb hexyl (5 µm)	Acetonitrile or methanol with phosphate or acetate buffer.	UV or fluorescence	Analysis in plasma using solid phase column extraction.	[150]
10 cm × 4.6 mm Brownlee RP-8 (5 µm)	100% methanol-aqueous phosphate buffer (pH 3) (1:1) (1 mL/min.).	460 nm	Post column derivatization of the drug using N-methyl 9-chloro-acridinum triflate	[151]

Table 6 (continued)

High Performance Liquid Chromatography Systems Used in the Analysis of Procainamide Hydrochloride

5 or 20 cm × 4 mm TSK Gel SCX, Nucleosil 5 SA or TSK ODS 120T (each 5 μm)	–	Photometric or fluorimetric detection.	Analysis of drug in tissue homogenates by HPLC with direct injection and column switching.	[152]
5 cm × 4.6 mm, normal and tertiary butyl alkyl bonded phases (5 μm)	Aqueous 4% triethylamine (pH 4.8) methanol (19:1) containing 10 mM KH_2PO_4	254 nm	Comparison and study of normal and tertiary butyl alkyl bonded phases for RPHPLC.	[153]
22 cm × 4.6 cm RP-18 (Spheri-5) (5 μm)	Methanol–water containing sodium dodecyl sulfate and NaH_2PO_4 (2 mL/min.).	UV	Analysis of the drug in biological fluids by direct injection, multi-dimensional HPLC with a micellar cleanup and RPC.	[154]

Table 6 (continued)

High Performance Liquid Chromatography Systems Used in the Analysis of Procainamide Hydrochloride

5 cm × 4 mm of Nucleosil 5SA (5 μm)	0.1 M – phosphate buffer (pH 3) containing 30% of acetonitrile.	280 nm	Analysis, by direct injection of whole blood sample of the drug and its N-acetyl metabolite.	[155]
15 cm × 4.5 mL IBM phenyl (5 μm.).	75m M- Acetate buffer (pH 4.3)-acetonitrile (20:3) (1 mL, min.).	270 nm	Analysis of the drug and three of its metabolites in serum and urine by RPHPLC.	[156]
10 cm × 1 mm of Spherisorb C_{18} (3 μm.).	0.025 M- Phosphate buffer (pH 3)-acetonitrile (9:1) (80 to 100 μL/min.).	UV	Analysis of the drug and its metabolite in serum using Micro-bore HPLC for drug monitoring and toxicology.	[157]
15 cm × 4.8 mm stainless steel, packed with Nucleosil C_{18} (5 μm.).	Methanol-acetic acid-triethylamine-water (40:2:1:157) (2 mL/min.).	335 nm (extraction at 290 nm.).	Analysis of the drug and its metabolite in human blood plasma.	[158]

Table 6 (continued)

High Performance Liquid Chromatography Systems Used in the Analysis of Procainamide Hydrochloride

25 cm × 4.6 mm of CN-type material (5 μm) operated at 40°C.	Acetonitrile-methanol-buffer (60:7:3), buffer is 10 mM phosphate – 0.5 mM-triethylamine, pH was adjusted to 7.1 (2 mL/min.).	205 nm	Simultaneous determination of the drug and its metabolite with other antiarrthymics in serum.	[159]
μBondapak ODS	Aqueous 35% methanol - 0.06M acetate buffer (pH 4.5).	220 nm	Analysis of the drug and its N-acetyl metabolite in human serum.	[160]
5 cm × 5 mm of Protein-coated ODS (20 to 32 μm).	Acetonitrile in 0.1 M phosphate buffer	UV or fluorescence	Determination of the drug by direct injection analysis of plasma sample.	[161]
Partisil-10 ODS	Methanol - 0.15 M $NH_4H_2PO_4$ (pH 4.5) (7:18) containing 0.1% of propionic acid.	245 nm and 280 nm	Analysis of the drug and its main metabolites in blood serum.	[162]

Table 6 (continued)

High Performance Liquid Chromatography Systems Used in the Analysis of Procainamide Hydrochloride

Nucleosil 5 SA (5 μm)	–	–	Analysis of the drug by direct injection of whole blood samples.	[163]
–	Five slightly different mobile phases.	–	Analysis of Procainamide, its N-acetyl derivative and other drugs by RPHPLC.	[164]
–	–	280 nm	Analysis of the drug and its N-acetyl metabolite in serum.	[165]
μBondapak C_{18} column	Acetonitrile in 0.1 mol/L potassium phosphate buffer pH 4 (9.75/90.25 by vol.)	–	Analysis of the drug and its N-acetyl metabolite in serum.	[166]

Table 6 (continued)

High Performance Liquid Chromatography Systems Used in the Analysis of Procainamide Hydrochloride

C_8 Reversed phase column.	80% Phosphate buffer (25 mmol/L, pH 3.4)-20% organic (acetonitrile: methanol, 2:3) and then changed to 30% phosphate 70% organic at 20 min after injection.	212 nm	Analysis of antiarrhythmic drugs and their metabolites from serum extracts.	[167]
Octyl-bonded reversed phase column.	Acetonitrile-phosphate buffer.	Ultraviolet or phosphorescence.	Analysis of antidysrhythmic drugs in serum.	[168]
μBondapak C_{18} column	Triethylamine-acetic acid or N-ethyl morpholine-acetic acid at pH 5.3.	–	Analysis of Procainamide and its N-acetyl metabolite in serum using a volatile mobile phase.	[169]

Table 6 (continued)

High Performance Liquid Chromatography Systems Used in the Analysis of Procainamide Hydrochloride

30 cm × 4 mm i.d. of alkylphenyl column	Acetonitrile and phosphate buffer pH 6.6 (60:40) (2 mL/min).	280 nm	Determination of the drug and its N-acetyl derivative in biological fluids.	[170]
15 cm × 3.9 mm i.d. of a Symmetry Shield RP8 and RP8 Sentry guard column (2 cm × 3.9 mm i.d.)	25 mM potassium phosphate of pH 7 – methanol (7:13) (1 mL/min).	254 nm	Analysis of the drug in serum. A novel polymeric reversed phase sorbant for solid phase extraction.	[171]
25 cm × 4.6 mm i.d., (5 μm) C_{18} ultrasphere or a 5 μm Supelcosil ABZ plus C_{18} column.	Methanol-0.005 M – Tris of pH 7 (1:4)	–	Post column immuno-detection following conditioning of the HPLC mobile phase by an on line ion exchange extraction.	[172]

Table 6 (continued)

High Performance Liquid Chromatography Systems Used in the Analysis of Procainamide Hydrochloride

15 cm × 4.6 mm i.d. of Synchropak SCD-100 (5 μm)	25 mM-Ammonium sulfate (pH 7.3) containing 24% of acetonitrile (1.3 mL/min).	230 nm.	Analysis of Procainamide in serum. The drug did not interfere with the analysis of tocainide	[173]
0.5 m × 2.6 mm i.d. of ODS-SIL-X-1	Acetonitrile: phosphate buffer (10:90). (2 mL/min).	205 nm	Analysis of the drug and lidocaine in serum.	[174]
30 cm × 4 mm i.d. of μBondapak-C_{18} (10 μm)	Isocratic 4 gm sodium acetate 1000 mL H_2O, 40 mL acetic acid, 50 mL acetonitrile (2 mL/min).	254 nm	Quantitation of the drug and its N–acetyl metabolites in plasma.	[175]
30 cm × 3.9 mm i.d. of μBondapak-C_{18}	100 g/L acetonitrile solution in sodium acetate buffer (0.75 mole/liter, pH 3.4) (2 mL/min).	280 nm	Measurement of the drug and its N–acetyl metabolites in serum.	[176]

Table 6 (continued)

High Performance Liquid Chromatography Systems Used in the Analysis of Procainamide Hydrochloride

30 cm column of μBondapak-C_{18} (reversed plasma).	Methanol in water. 400 mL/L with glacial acetic acid, 10 mL/L (2 mL/min) or methanol/H_2O 40/60 by volume and 1/99 by volume of acetic acid (2 mL/min).	280 nm	Analysis of the drug and its N-acetyl derivative in plasma or serum.	[177]
Stainless steel column (30 cm × 4 mm i.d.) of bonded phase packing.	Ammonium hydroxide: water (10:1) (5 mL/min).	280 nm	Rapid analysis of the drug and its N-acetyl derivative in human plasma.	[178]
25 cm × 2.1 mm of Zorbax-SIL silica column.	100 mL of methanol, 2 mL water, and 0.1 mL of morpholine (1.9 mL/min).	254 nm	Determination of plasma Procainamide and its N-acetyl derivative concentration.	[179]
μBondapak CN	Acetate buffer solution of pH 4.5	254 nm	Fluoro immunoassay and HPLC compared	[180]

4.5.7 Soap Chromatography

Lam *et al* reported the determination of Procainamide and N-acetyl Procainamide by using soap chromatography [181].

4.5.8 Capillary Electrophoresis

Gillott *et al* separated Procainamide and other pharmaceutical bases from neutral and acidic components by electrochromatography (CEC) [182]. The analysis of the strong pharmaceutical bases by CEC using several commercially available reversed phase materials was achieved by the addition of competing bases such as triethylamine or triethanolamine to the mobile phase at pH 2.5. CEC was performed as described previously by Euerby *et al* [183] with a mobile phase comprising acetonitrile mixed with various amounts of 0.05 M buffer, water and modifier, and detection at 214 nm. This method permitted the separation of acids, neutrals, and bases in a single CEC run using triethanolamine as the competing base. Efficiencies up to 510,000 plates/m with acceptable peak symmetry were achieved for Procainamide and other bases on Hypersil C_{18}, C_8, and phenyl reversed-phase packing materials.

The CEC separation of Procainamide and other pharmaceutical bases was investigated by Gillott *et al* on a commercially available silica stationary phase using aqueous mobile phases [184]. The effect of mobile phase composition, buffer pH, applied voltage, and buffer anion on the retention behavior of these bases was studied. Promising chromatography was obtained at pH 7.8, but was later found to be irreproducible. Successful and reproducible chromatography of the drug and other bases was achieved at pH 2.3. The authors reported that the addition of mobile phase additives such as TEA-phosphate at low pH values resulted in excellent CEC analysis of the bases on reversed-phase packing materials. The same approach was applied to the analysis of the drug and the other bases on silica phases in order to improve peak shape. Excellent chromatography was claimed for the analysis of strong pharmaceutical bases. Investigations have shown that the CEC separation of a range of pharmaceutical bases can routinely be achieved with excellent peak shapes, and with peak efficiencies as high as 320,000 plates/m.

Pesakhovych used 1,3,5-trinitrobenzene as electroosmosis and rheophoresis indicators in the analysis of electrophoretic spectra [185]. In experiments on the electrophoresis of Procainamide and other N-containing drugs, with tetraethylammonium iodide as a standard, it was shown than 1,3,5-trinitrobenzene and blue dextran were equally effective as electroosmosis and rheophoresis indicators. However, the former was preferred, since like the drug and standard it gave a circular zone and the dextran gave a comet-like zone. Reagents for development of the various zones were given.

Thomas *et al* discussed the application of a high performance capillary electrophoresis (HPCE) for ligand binding analysis of drug-protein systems [186]. The metabolites of Procainamide were studied as model compounds with respect to their irreversible binding to hemoglobin and histone proteins. Using Hughes-Klotz analysis, free-zone HPCE was shown to give comparable binding constants to those obtained by flow-injection analysis with electrochemical detection. In addition to analyzing the concentration of free ligand to deduce the concentration of bound ligand, the peak profile of free versus bound protein was studied in order to obtain relevant information about the specific nature of the bound product. The method is applicable in the analysis of samples of limited volumes, and can be applied in a single analysis to the determination of ligands able to bind to several different macromolecules.

Vargas *et al* directly determined Procainamide and N-acetyl-Procainamide by a capillary zone electrophoresis method, applying this to pharmaceutical formulations and to urine [187]. Samples were injected hydrodynamically for 5 seconds without prior treatment onto a fused-silica capillary (43.5 cm × 75 μm i.d.; 35.9 cm to the detector), to which a voltage of 10 kV was applied. The background electrolyte was 50 mM-phosphate buffer, and detection was at 200 nm. Detection limits were 1.2 μg/mL for Procainamide, and the calibration graph was linear up to 200 μg/L.

Shihabi used a capillary electrophoresis method for the analysis of serum Procainamide, based on acetonitrile stacking [188]. Serum containing an internal standard was deprotonized with acetonitrile and used to fill 12% of the capillary volume. The running buffer contained triethanolamine, 2-(N-cyclohexylamino)ethanesulfonic acid, and 20% isopropanol of pH 8.2. Procainamide, N-acetyl-Procainamide, and the internal standard (quinine) were separated in seven minutes. Results compared well with those obtained using an immunoassay.

4.6 Microdialysis

Sarre *et al* adopted a microdialysis method for determination of the in plasma *in vitro* protein binding of Procainamide and some other drugs [104]. An artificial blood vessel was constructed and filled with spiked plasma circulating at 15 mL/minute (the flow rate of human blood) at 37°C. The microdialysis probe (16 mm membrane 20,000 MW cut off) was placed in the vessel and perfused with 0.9% sodium chloride at 5 μL/min. Dialysates were analyzed by HPLC.

4.7 Immunoassay Methods

Mojaverian and Chase reported the production and the characterization of specific antibodies for use in the radioimmunoassay of Procainamide, and the measurement of the drug in human serum *in vitro* and *in vivo* [191]. The serum concentration of Procainamide was determined in a 26-year-old male subject after an IV dose of the drug (461 mg/30 mL), given during 10 minutes by the use of infusion pump. Results showed that the assay allows the direct measurement of the drug in a 0.1 mL aliquot of diluted serum. As little as 1 ng of the drug per milliliter of serum can be detected.

Bach and Larsen studied the stability of standard curves prepared for enzyme-multiplied immunoassay technique (EMIT) homogenous enzyme-immunoassay kits stored at room temperature after reconstitution, and reported the determination of Procainamide by this method [192].

Four fluorescence-polarization immunoassay methods for the monitoring of Procainamide and other drugs were evaluated by Haver *et al* [193]. Serum was analyzed for the drug with a fluorescence-polarization immunoassay kit (Roche) on the Cobas Bio centrifugal analyzer in comparison with the Abbot TDx and EMIT (Syva) methods and gas chromatography.

Walberg and Wan determined the drug and its N-acetyl metabolite by EMIT, and reported their clinical evaluation of the EMIT Procainamide and N-acetyl-Procainamide assay [194]. The basic principles of EMIT-type enzyme immunoassays, and the determination of Procainamide and its acetyl derivative by EMIT, was reported by Spitzbarth [195]. Shaw and McHan reported their adaptation of the EMIT procedures for maximum cost effectiveness to two different centrifugal analyzer systems, and reported the determination of Procainamide, in serum, with EMIT kits on centrifugal analyzers [196].

Zvaigzne reported the statistical analysis of the stability of the standard curves for some Syva EMIT assays [197]. Separate EMIT determinations for Procainamide were used to examine the stability of calibration graphs over periods of up to 90 days. The calibration was performed with the Syva EMIT Autolab 5000 system, and different mathematical algorithms were re-written in the form of a linear equation. The calibration graph was found to be stable for 89 days for Procainamide.

A more sensitive enzyme-multiplied immunoassay technique for the quantitative determination of Procainamide, and N-acetyl-Procainamide, in plasma, serum, and urine, was reported by Henry and Dhruv [198]. The drug and its derivative were determined by a commercial EMIT method (Syva Co), modified to permit automated analysis of about 100 samples per day in a working range of 0.1 to 2.0 μg/mL. A Gilford 203S continuous-flow system was used in conjunction with a Beckman Model DU Spectrophotometer.

An enzyme immunoassay, liquid chromatographic, and fluorimetric assay methods were compared and used by Pape for the determination of Procainamide in serum [92].

The evaluation of the very rapid enzyme multiplied immunoassay (EMIT Qst) methods for measuring serum Procainamide and N-acetyl-Procainamide concentrations was reported by Carnes and Coyle [133]. To validate the EMIT Qst method for Procainamide and N-acetyl-Procainamide by comparing these with the established high performance liquid chromatography assays, a study was conducted to examine the accuracy of the assay and the intra-day and inter-day precision. This was accomplished by measuring Procainamide and N-acetyl-Procainamide in spiked serum samples (2-15 mg/L) via both the EMIT Qst and HPLC methods. The correlation between the methods was then assessed in 39 samples from 20 patients taking Procainamide as well as concurrent medications. The results indicated that the EMIT Qst assays are rapid, accurate, and precise for routine clinical measurements of Procainamide and N-acetyl-Procainamide. Both the EMIT Qst and HPLC methods appeared specific in the presence of a number of other medications.

Wesley and Lasky reported an HPLC method that was correlated with an enzyme immunoassay technique (EMIT) for the determination of Procainamide and N-acetyl-Procainamide in blood serum [135]. Griffith *et al* described a reversed phase HPLC assay method for the analysis of Procainamide and its N-acetyl derivative, and compared this with an EMIT method for the assay of the drug and derivative in serum [137]. A comparison was also made by Gallaher *et al* between a high performance liquid chromatography and enzyme-multiplied immunoassay for the analysis of Procainamide in serum [139].

Mackichan *et al* compared a fluoroimmunoassay method for Procainamide and N-acetyl-Procainamide with a high pressure liquid chromatography method [180]. Al Hakiem *et al* developed a fluoroimmunoassay method for the determination of individual or combined levels of Procainamide and N-acetyl Procainamide in serum [199].

Sonsalla *et al.* evaluated a TD_X fluorescence polarization immunoassay method for Procainamide and N-acetyl-Procainamide [200]. The method was performed by using antibodies to Procainamide raised in rabbits and antibodies to N-acetyl-Procainamide raised in sheep. Fluorescein was used as the fluorescent label. The method was used for the analysis of serum and plasma. Lam and Watkins conducted a fluorescence polarization immunoassay method for the analysis of the drug and its N-acetyl derivative using Cobas-FP reagents [201].

5. Stability

Procainamide hydrochloride is more stable in neutral solutions such as sodium chloride, than in acidic solutions such as glucose. However, patients requiring intravenous Procainamide often have heart failure and cannot tolerate the sodium load associated with sodium chloride injections. The stability of Procainamide in glucose 5% injection is improved by neutralizing the glucose injection using sodium bicarbonate, or by storing the admixture at 5°C. The concentration of Procainamide remained above 90% of the initial concentration for 24 hours if the glucose injection was first neutralized, and this was considered more practical than refrigeration if extended stability was required [10,11].

The compound formed by mixing Procainamide hydrochloride with glucose 5% infusion was shown to be a mixture of α- and β-glucosylamines [12]. About 10 to 15% of Procainamide was lost in this way after 10 hours at room temperature [11].

Procainamide hydrochloride oral preparations should be stored at room temperature in tight containers, and exposure to temperatures warmer than 40°C should be avoided. Procainamide hydrochloride injections are colorless or may turn slightly yellow on standing. Solutions of Procainamide that are darker than light amber or otherwise discolored should not be used. Although Procainamide hydrochloride injection may be stored at room temperature (10-27°C), refrigeration retards oxidation and associated development of color. When the injection formulation is diluted with 0.9% sodium chloride injection or sterile water for injection, solutions containing 2-4 mg/mL are stable for 24 hours at room temperature or for 7 days at 2-8°C.

Procainamide hydrochloride injection has been reported to be physically incompatible with some drugs, but the compatibility depends on several factors such as concentration of the drugs, specific diluents used, resulting pH, and temperature. Specialized references should be consulted for specific compatibility information [13].

6. Drug Metabolism and Pharmacokinetics

6.1 Adsorption

The absorption of Procainamide hydrochloride is rapid through the oral or the intramuscular route, and is immediate through the intravenous route [9]. The drug is readily absorbed from the gastrointestinal tract of healthy subjects, with a short half-life of 2.4-4 hours [15]. Approximately 75% to 95% of the dose from non-sustained-release or sustained-release forms is absorbed orally, although absorption of less than 50% is observed in small percentage of patients [38]. The bioavailability of sustained-release Procainamide is estimated to be 68% [31].

The rate of the drug release is controlled in the Procanbid® dosage form so that absorption is sustained throughout the 12-hour dosing intervals [39]. The rate of Procainamide absorption is similar in men and women, and in black and Caucasian patients [22]. The effect of food on the absorption characteristics of Procainamide hydrochloride was assessed after oral administration of the drug given to patients in the fasting and postprandial states. Serum concentration-time curves showed no significant differences in peak serum levels of the drug, in the time the peak value was reached, or in the area under the serum concentration-time curves, indicating the total amount of Procainamide HCl present in the serum. Thus, taking an oral dose of the drug with food does not significantly alter the bioavailability of Procainamide [40]. A recent study [39] showed that when Procanbid® is administered with a high fat meal, the extent of absorption of the Procainamide was increased by approximately 20%.

6.2 Distribution

Procainamide is rapidly distributed into the cerebrospinal fluid, liver, spleen, kidney, lungs, muscles, brain, and heart. The apparent volume of distribution (V_d) of the drug at steady state is 1.48 to 4.3 L/kg, and the V_d for the central compartment ranges from 0.1 to 0.9 L/kg [12,19,28,38,41]. The distribution volume is decreased in patients with heart failure. Studies using radiolabeled Procainamide indicated that 14% to 23% of the drug is bound to plasma proteins at therapeutic plasma concentrations. Procainamide crosses the placenta, but the extent to which it does has not been well characterized [12,39]. Procainamide and N-acetyl-Procainamide is distributed into milk, and can be absorbed by nursing infants [12].

Following oral doses of 15 to 22.5 mg/kg, arrhythmias were suppressed within 1 to 2.5 hr [14]. There is a considerable inter-subject variation in plasma concentrations after dosing. The therapeutic effect of Procainamide has been correlated with plasma concentrations of about 4 to 10 μg/mL, with progressively severe toxicity being noted at concentrations above 12 μg/mL [12,15]. Several reports [16-19] indicate that suppression of arrhythmias occurs at a mean plasma level of 7.5 μg/mL. Optimal plasma or serum concentrations outside this range were reported for some individuals, and some data [20] indicate that plasma concentrations as high as 10 to 15 μg/mL are required for effective therapy.

Twice daily administration of Procanbid® to healthy subjects produced peak, trough, and mean plasma Procainamide levels similar to those achieved when Procan SR® was administered four times daily. Equivalent daily doses of Procanbid® and Procan SR® in patients with frequent ventricular arrhythmias produced bioequivalent peak and steady-state mean Procainamide levels. However, corresponding minimum levels for Procanbid SR® are slightly lower than those for Procan SR®, but remain within the therapeutic range. A mean plasma Procainamide level of 4.6 μg/mL with mean peak and trough levels within the therapeutic range produced by twice daily administration of two 1000 mg Procanbid® tablets to patients with frequent ventricular arrhythmias [21]. Similar concentrations of Procainamide and N-acetyl-Procainamide are produced in black and Caucasian patients following administration of Procanbid® tablets [22].

In another study [23], a single oral dose of 1g attained peak plasma concentrations of 3.5 to 5.3 μg/mL of Procainamide (mean 4.2) in 1 to 2 hr. Concentrations of N-acetyl-Procainamide reached a peak of 0.6 to 2.1 μg/mL (mean 1.6) in 3 to 8 hr. Many reports [24-30] estimated a time to peak concentration within 2 to 4 hours for oral sustained-release. Peak levels of N-acetyl-Procainamide occurred with 1 to 8 hours following oral administration [23]. There was a suggestion that the bioavailability of sustained-release preparations may be reduced by 15% to 20% compared with conventional solid dosage formulations [31].

Following IV administration, peak levels are reached within 20 to 30 minutes and are maintained for 1 to 2 hours [20]. Peak plasma Procainamide concentrations after IM administration of the drug averaged 30% higher than after oral administration of the same dose. In one study [12] in healthy individuals, peak plasma Procainamide concentrations of 5-8.5 μg/mL were attained in 15-60 minutes, and plasma concentrations of 2-3 μg/mL persisted for 6 hours after a single 1 gm IM dose. In another study in healthy Chinese [32], the log time absorption was about half an hour. The peak time of Procainamide and its active metabolite, N-acetyl-Procainamide, occurred at about 1 and 3 hour, respectively, after dosing.

In one study [22] using Procanbid®, an AUC range of 19.1 to 24.5 μg × hr/mL and 17.6 to 21.5 μg × hr/mL for N-acetyl-Procainamide were estimated. According to the study, the AUC's for Procainamide and N-acetyl-Procainamide were similar in black and Caucasian patients. For women, the AUC was 28% greater than for men. This difference was reduced to 13% after normalization of individual AUC values for body surface area. The AUC for N-acetyl-Procainamide was the same for men and women.

Plasma concentrations of the drug may be elevated in certain disease, such as cardiac and renal insufficiency [12]. N-acetyl-Procainamide may represent a significant fraction of the total drug in circulation. The therapeutic effect has been correlated with plasma concentrations of about 5 to 30 μg/mL of combined Procainamide and N-acetyl-Procainamide. The latter may accumulate during chronic administration in fast acetylators, and in subjects with renal impairment [11]. The steady state plasma concentrations of Procainamide and metabolites were reported in ten subjects receiving the following maintenance treatment: Procainamide, 2.6-20.7 μg/mL (mean 10.6); N-acetyl-Procainamide, 4.6 to 43.6 μg/mL (mean 15.9); monodesethyl-N-acetyl-Procainamide, 0.4 to 11.9 μg/mL (mean 2.7) [33].

Following intravenous infusion of 84 to 374 mg/hr to 34 subjects, steady state plasma concentrations of 1.7 to 17 μg (mean 6.5) of Procainamide and 1.1 to 20 μg/mL (mean 5.7) of N-acetyl-Procainamide were reported [34]. Therapeutic monitoring of the drug in saliva (including details of studies on saliva-Procainamide concentrations) revealed variability in the saliva-plasma ratio, making it impossible to predict plasma concentration from salivary drug concentration. But since saliva concentrations seem to reflect the drug concentration at an active cardiac site, they may be clinically relevant [35]. Marked intra- and inter-individual variation of N-acetyl-Procainamide salivary concentrations were observed in the elderly. These results indicated that salivary concentrations would be of no clinical value for monitoring therapy [36].

6.3 Metabolism

Biotransformation through N-acetylation in the liver takes place in approximately half of the dose, with the acetylation rate being genetically determined [42]. The activity of the N-acetyltransferase enzyme has been linked to the efficacy and toxicity of Procainamide. Based on the activity of this enzyme, populations are divided into two groups; the so called slow acetylators and rapid acetylators. About 50% to 65% of Caucasians, Blacks, South Indians, and Mexicans are slow acetylators, while 80% to 90% of Eskimos, Japanese, and Chinese are rapid acetylators. The acetylator status alters the half-life and clearance of Procainamide, however the differences are not significant. The additive effects of these factors become significant when high doses are given to slow acetylators. Saturation of the acetylation mechanism may occur, leading to lower clearance and unexpected high plasma levels and may result in toxicity [43-45]. The major metabolite of Procainamide is N-acetyl-Procainamide (NAPA) which possesses similar anti-arrhythmic potency to the parent drug. NAPA exists in a ratio of 0.5 to 2.4 to the Procainamide in plasma after chronic administration. Slow acetylators metabolize approximately 16% to 21% of Procainamide to the N-acetyl derivative while rapid acetylators metabolize 24% to 33% [39,42,46].

Other metabolites include desethyl-N-acetyl-Procainamide and desethyl-Procainamide. Although desethyl-N-acetyl-Procainamide was found to be a metabolite of N-acetyl-Procainamide, it is suggested that the biotransformation of Procainamide to this metabolite is not *via* the N-acetyl-Procainamide intermediate. As illustrated in Figure 12, it has been suggested instead that the principle metabolic route of desethyl-N-acetyl-Procainamide from Procainamide occurs by initial dealkylation, to form *p*-amino-N-[2-(ethylamino)ethyl]-benzamide. This intermediate has not, however, been identified [47]. No activity has been reported for desethyl-N-acetyl-Procainamide.

6.4 Elimination

The drug and its metabolites are excreted mainly by the kidney, but excretion in breast milk and bile has been reported. Both Procainamide and NAPA are excreted in human breast milk and can be absorbed by a nursing infant [39]. Approximately 40% to 70% of Procainamide is excreted in urine as the unchanged drug, and the remainder was constituted by N-acetyl-Procainamide and other metabolites (such as mono-desethyl-Procainamide and monodesethyl-N-acetyl-Procainamide). Renal clearance of 150 to 600 mL/min was observed [23,28,41,46,48-52].

Excretion of N-acetyl-Procainamide may be a function of acetylator phenotype. The major metabolite is excreted unchanged in a range of about 15% to 80% [23,28,39,41,46,48,49,52]. A renal clearance range of 285 to 529 mL/min has also been reported for Procainamide [32]. Renal excretion of unchanged form accounted for a mean of 56% of Procainamide elimination. The recovery of NAPA in urine was about 19% [32].

Figure 12. Metabolism of Procainamide.

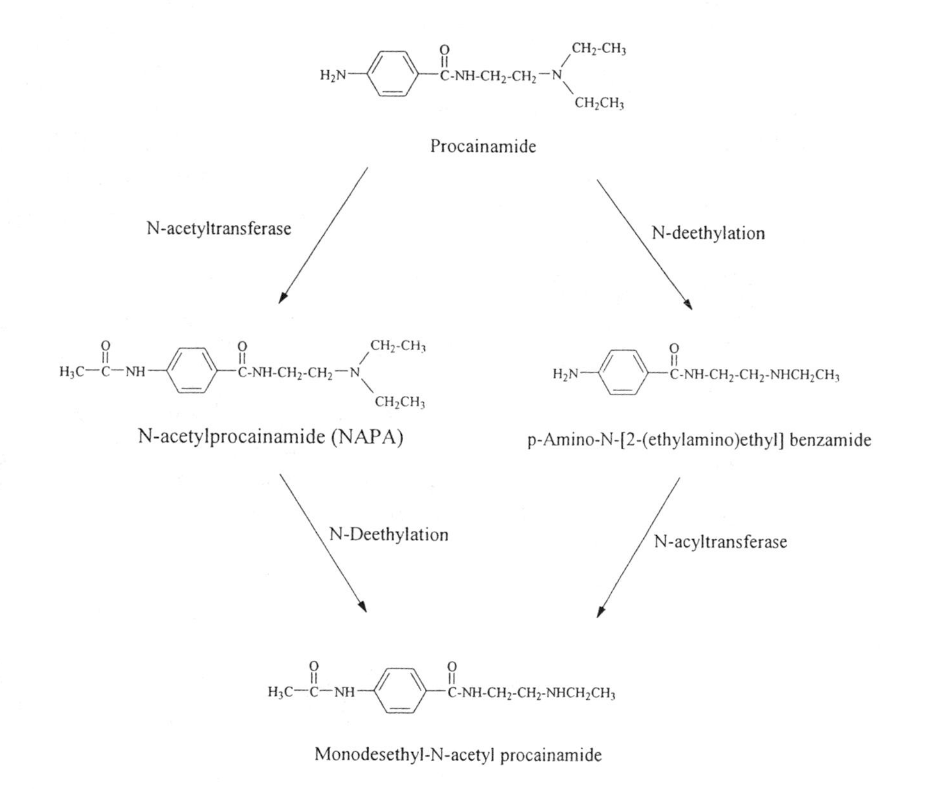

One study [53] indicated a bile/plasma ratio of less than 1.0, and that the total bile excretion and bile clearance is probably insignificant. Elimination half-lives ranged from 2.5 to 8.0 hr [19,23,28,38,41,45,48,54]. An average half-life of 8.52 ± 3.58 has also been reported [56]. The half-life is prolonged in elderly patients, renal failure, and severe heart failure [39,53] but unaltered in patients with myocardial infarction or mild congestive heart failure [56]. The half-life of N-acetyl-Procainamide is 5 to 9 hr, and may be prolonged in renal failure [51,57]. Clinical data from hospital inpatients receiving Procainamide for the treatment of various cardiac arrhythmias showed that mild congestive heart failure was associated with only minor effects on Procainamide kinetics, although this did not discount the possibility of a greater influence of more severe heart failure [31].

Extra-corporeal elimination was studied using such techniques as hemodialysis, peritoneal dialysis, and hemofiltration. Hemodialysis (HD) removes significant amounts of the drug and its active metabolite, and can reduce the half-life of Procainamide by about one-half of the non-dialysis value. Procainamide has a dialysis clearance of 65 to 75 mL/min, while the value for N-acetyl-Procainamide is 48 to 54 mL/min. [50-52,58,59]. In a case report of a patient with Procainamide toxicity and end-stage renal disease, the use of combined HD, hemoperfusion (HP), and continuous ambulatory peritoneal dialysis (CAPD) was effective in the plasma clearance of Procainamide. HD was more effective than HP in clearing N-acetyl-Procainamide. Combination of HD/HP, however, increased substantially the clearance of Procainamide and N-acetyl-Procainamide attainable by either modality alone. Four 4-hour sessions of HD/HP removed about 34% of the initial total body burden of Procainamide. About 31% of the initial N-acetyl-Procainamide burden was removed by the first 4-hour session of combined HD/HP. The amount of Procainamide and N-acetyl-Procainamide removed by CAPD was about 19% to 24%, respectively, of the total amount removed by all treatment modalities. The half-life of Procainamide and N-acetyl-Procainamide was 38 hr and 39 hr, respectively [59].

Peritoneal dialysis of the drug is minimal [60-62]. Procainamide is also dialyzable by hemoperfusion [58,59]. During hemoperfusion in a patient with toxicity and renal failure, the clearances of Procainamide and N-acetyl-Procainamide were 73 to 75 mL/min for both agents. Combination of HD and HP resulted in a N-acetyl-Procainamide clearance almost 3 times greater than that with dialysis alone [58]. Dialysis using hemofiltration is also reported for Procainamide [63].

7. Toxicity

Toxicity generally occurs at plasma or serum levels above 8 to 16 μg/mL [19], and fatalities are associated with concentration exceeding 20 μg/mL. As little as 200 mg IV can be fatal. In four fatalities attributed to Procainamide intoxication in subjects receiving therapeutic doses, plasma-Procainamide concentrations shortly before death ranged from 17.6 to 25.2 μg/mL [19]. Postmortem Procainamide concentrations of 114 μg/mL blood, 293 μg/g liver, and 556 μg/mL urine were reported in a fatality attributed to Procainamide [37].

Overdosage of Procainamide has produced hypotension, junctional tachycardia, oliguria, lethargy, confusion, nausea, and vomiting [19].

8. Drug Interactions

8.1 H_2-Receptor Antagonists

Concomitant administration of Procainamide and cimetidine may result in increased plasma Procainamide and N-acetyl-Procainamide (NAPA) concentrations, and subsequent toxicity. This interaction may be more marked in geriatric patients and in patients with renal impairment since such patients eliminate Procainamide, NAPA, and cimetidine more slowly. Cimetidine decreases the renal clearance of Procainamide and NAPA, however, additional mechanisms may also contribute to this interaction.

Limited evidence suggests that ranitidine may increase plasma concentrations of Procainamide and NAPA, but to a lesser extent than cimetidine. The precise mechanisms for this interaction are complex and are not fully understood. Evidence to date suggests that famotidine does not substantially interact with Procainamide. Caution should be exercised when either cimetidine or ranitidine is administered concomitantly with Procainamide, particularly in geriatric patients and patients with renal impairment. The patient and plasma Procainamide concentrations should be monitored closely and Procainamide dosage adjusted accordingly [12].

8.2 Neuromuscular Blocking Agents

Procainamide may potentiate the effects of both non-depolarizing and depolarizing skeletal muscle relaxants, such as gallamine triethiodide, metocurine iodide, pancuronium bromide, succinylchloine chloride, and tubocurarine chloride. Although the clinical significance of this interaction has not been established, Procainamide should be used with caution in conjunction with neuromuscular blocking agents [12].

8.3 Anticholinesterase and Anticholinergic Agents

Procainamide should be used with caution (if at all) in patients with Myasthenia gravis, and the dose of anticholinesterase drugs such as neostigmine and pyridostigmine may have to be increased. Theoretically, the anticholinergic effect of Procainamide may be additive with anticholinergic drugs [12].

8.4 Cardiovascular Drugs

Since Procainamide may reduce blood pressure, patients receiving parenteral hypotensive drugs and Procainamide, or in high oral doses, should be observed for possible additive hypotensive effects [12].

When Procainamide is administered with other anti-arrhythmic drugs such as lidocaine, phenytoin, propranolol or quinidine, the cardiac effects may be additive or antagonistic and toxic effects may be additive [11].

Concomitant use of Procainamide and amiodarone may result in increased plasma Procainamide and N-acetyl-Procainamide concentrations and subsequent toxicity. In a limited number of patients receiving 2-6 g of Procainamide hydrochloride daily, initiation of amiodarone hydrochloride (1200 mg daily for 5-7 days and then 600 mg daily) increased plasma Procainamide and NAPA concentrations by about 55% and 33%, respectively, during the first week of amiodarone therapy. The exact mechanism(s) has not been elucidated, but it has been suggested that amiodarone may decrease the renal clearance of Procainamide and NAPA and/or inhibit the hepatic metabolism of Procainamide. In addition to a pharmacokinetic interaction, additive electrophysiologic effects (including increased QTC and QRS intervals) occur during concomitant use, adverse electrophysiologic effects (e.g., acceleration of ventricular tachycardia) may also occur. Pending further accumulation of data, it is recommended that Procainamide dosage be reduced by 20-33% when amiodarone therapy is initiated in patients currently receiving Procainamide, or that the Procainamide therapy be discontinued [12].

Acknowledgement

The authors wish to thank Mr. Tanvir A. Butt, Department of Pharmaceutical Chemistry, College of Pharmacy, King Saud University, for typing this profile.

References

1. ***The Merck Index***, 12th edition, Merck and Co., Inc. Rahway, N.J. 1989, p. 1331.

2. ***The United States Pharmacopoeia***, **24**, United States Pharmacopoeial Convention, Inc., Rockville, MD, 2000, p. 1397; [a] ***USP 20***, 1980 p. 659.

3. ***The British Pharmacopoeia, 1988***, Her Majesty's Stationary Office London (1988), p. 464; [a] ***Ibid 1993***, Volume 1, p. 543.

4. F.R. Mautz, *J. Thorac. Surg.*, **5**, 612 (1936).

5. L.C. Mark, H.J. Kayden, J.M. Steele, J.R. Cooper, I. Berlin, E.A. Ovenstine, and B.B. Brodie, *J. Pharmacol. Exp. Therap.*, **102**, 5 (1951).

6. A.G. Gilman, L.S. Goodman, and A. Gilman, eds. "***Goodman and Gilman's The Pharmacological Basis of Therapeutics***," 6th edn., Macmillan Publishing Co., Inc. New York, 1980, p. 774.

7. R.B. Poet and H. Kadin, ***Analytical Profiles of Drug Substances***", K. Florey, ed., Volume 4, Academic Press, New York, 1975, p. 333.

8. ***The Pharmaceutical Codex***, The Pharmaceutical Press, London, 1979, p. 739.

9. ***Clarke's Isolation and Identification of Drugs in Pharmaceutical, Body Fluids and Post-Mortem Material***, 2nd edn., A.C. Moffat, ed., The Pharmaceutical Press, London, 1986, p. 924; [a] ***Ibid*** 1st edn., 1979, Volume I, p. 512.

10. G.G. Raymond, M.T. Reed, J.R. Teagarden, K. Story, and C.W. Geberbauer, *Am. J. Hosp. Pharm.*, **45**, 2513 (1988).

11. USP DI 1996, ***Drug Information for the Health Care Professional***, 16th, edn., Volume 1, 1996, p. 2472.

12. A. Sianipar, J.E. Parkin, and V.B. Sunderland, *J. Pharm. Pharmacol.*, **46**: 951 (1994).

13. AHFS 96, Drug Information, American Hospital Formulary Services, American Society of Health System Pharmacists, Bethesda, USA, 1996, p., 1197.

14. R.H. Levy, G.O. Gey, and R.A. Bruce, *J. Pharm. Sci.*, **63**, 1958 (1974).

15. ***Martindale, The Extra Pharmacopoeia***, 29th edn., J.E.F. Reynolds, ed., The Pharmaceutical Press, London, 1989, p. 933.

16. Anon, *Med. Lett. Drugs Therap.*, **31**, 35 (1989).

17. G.O. Gey, R.H. Levy, L. Fisher G. Pettet, and R.A. Bruce, *Ann. Intern. Med.*, **80**, 718 (1974).

18. E.G. Giardina, R.H. Heissenbuttel, and J.T. Bigger Jr., *Ann. Intern. Med.*, **78**, 183 (1973).

19. J.K. Weser and S.W. Klein, *J. Am. Med. Ass.*, **215**, 1454 (1971).

20. P. Chice *et al*, *Arch. Mal. Coeur.*, **865**, 797 (1972).

21. B.B. Yang, R.B. Abel, A.C. Uprichard, J.A. SmiTheraps, and S.T. Forgue, *J. Clin. Pharmacol.*, **36**, 623 (1996).

22. J.R. Koup, R.B. Abel, J.A. Smitheraps M.A. Eldon, and T.M. de Vries, *Therap. Drug Monit*, **20**, 73 (1998).

23. E.G. Giardina, J. Dreyfuss, J.T. Bigger, J.M. Shaw, and E.C. Schrieber, *Clin. Pharmacol. Therap.*, **19**, 339 (1976).

24. B.A. Baker, J.R. Reynolds, L. Gleckel, E. A'Zary, and M.M. Bodenheimer, *Clin. Pharm.*, **7**, 135 (1988).

25. D.E. Hilleman, A.J. Patterson, S.M. Mohiuddin, B.G. Ortmeier, and C.J. Destache, *Drug Intell. Clin. Pharm.* **22**, 554 (1988).

26. E.L. Michelson and L.S. Dreifus, *Vasc. Med.*, **2**, 155 (1984).

27. J. Birkhead, T. Evans, P. Mumford, E. Martinez, and D. Jewitt, *Br. Heart J.*, **38**, 77 (1976).

28. C. Graffner, G. Johnsson, and J. Sjogren, *Clin. Pharmacol. Therap.*, **17**, 414 (1975).

29. E. Karlsson, *Eur. J. Clin. Pharm.*, **6**, 245 (1973).

30. D. Fremstad, S. Dahl, S. Jacobsen, P.K. Lunde, K.J. Nadland, A.A. Marthinsen, T. Waaler, and KH Landmark, *Eur. J. Clin. Pharm.*, **6**, 251 (1973).

31. T.H. Grasela and L.B. Sheiner, *Clin. Pharmacokinet.*, **9**, 545 (1984).

32. M.Y. Lai, F.M. Jiang, and P.D. Leecha, *Taiwan Yaw Hsueh Tsa Chik*, **39**, 167 (1987).

33. T.I. Ruo, Y. Morita, A.J. Atkinson, T. Henthorn, and J.P. Thenot, *J. Pharmacol. Exp. Therap.*, **216**, 357 (1981).

34. J.J. Lima, A.L. Goldfarb, D.R. Conti, L.H. Golden, B.L. Bascomb, G.M. Benedetti, and W.J. Jusco, *Am. J. Cardiol*, **43**, 98 (1979).

35. M. Danhof and D.D. Breimer, *Clin. Pharmacokinet*, **3**, 39, (1978).

36. R.L. Galeazzi, C. Omer-Amberg, and G. Karlaganis, *Clin. Pharmacol. Therap.*, **29**, 440 (1981).

37. L. Kopjak and T.A. Jennison, *Bull. Inst. Ass. Fornes, Toxicol.*, **12**, 12 (1976).

38. J. Koch-Weser, *Ann. N.Y. Acad. Sci.*, **179**, 370 (1971).

39. *Product Information*: Procanbid®, Procainamide. Parke-Davis, Morris Plains, N.J., 1998.

40. W.D. McKnight and M.L. Murphy, *South. Med. J.*, **69**, 851 (1976).

41. K.J. Simons, R.H. Levy, R.E. Cutler, G.T. Christopher, and A. Lindner. *Res. Commun. Chem. Pathol. Pharmacol.*, **11**, 173 (1975).

42. J. Elson, J.M. Strong, W.K. Lee, and A.J. Atkinson, *Clin. Pharmacol. Therap.*, **17**, 134 (1975).

43. D.E. Drayer and M.M. Reidenberg, *Clin. Pharmacol. Therap.*, **22**, 251 (1977).

44. W.J. Tilstone, D.H. Lawson, W. Campbell, I. Hutton, and T.D. Lawrie, *Eur. J. Clin. Pharmacol.*, **14**, 261 (1978).

45. P. Hore, P. Bones, T. Rollinson, and H. Ikram, *Br. J. Clin. Pharmacol.*, **8**, 267 (1979).

46. M.M. Reidenberg, D.E. Drayer, M. Levy, and H. Warner, *Clin. Pharmacol. Therap.*, **17**, 722 (1975).

47. T.I. Ruo, J.P. Thenot, G.P. Stec, and A.J. Atkinson, *Therap. Drug Monit.* **3**, 231 (1981).

48. J. Dreyfuss, J.T. Bigger, A.I. Cohen, and E.C. Schreiber, *Clin. Pharmacol. Therap.*, **13**, 366 (1972).

49. E. Karlsson and L. Molin, *Acta Med. Scand.*, **197**, 299 (1975).

50. T.P. Gibson, E. Matusik, L.D. Nelson, and W.A. Briggs, *Clin. Pharmacol. Therap.*, **20**, 720 (1976).

51. T.P. Gibson, E. Matusik, and W.A. Briggs, *Clin. Pharmacol. Therap.*, **19**, 206 (1976).

52. T.P. Gibson, J. Matusik, E. Matusik, H.A. Nelson, J. Wilkinson, and W.A. Briggs, *Clin. Pharmacol. Therap.*, **17**, 395 (1975).

53. H.S. Weily and E. Genton, *Clin. Res.*, **19**, 169 (1971).

54. H.S. Weily and E. Genton, *Arch. Intern. Med.*, **130**, 366 (1972).

55. F. Jamali, R.S. Alballa, R. Mehvar, and C.H. Lemko, *Therap. Drug. Monit.*, **10**, 91 (1988).

56. K.M. Kessler, D.S. Kayden, D.M. Estes, P.L. Koslovskis, R. Siqueira, R.G. Trohman, A.R. Palomo, and R.J. Myerburg, *J. Am. Coll. Cardiol.*, **7**, 1131 (1986).

57. T.M. Ludden, M.H. Crawford, and G.T. Kennedy, *PharmacoTherapapy*, **5**, 11 (1985).

58. S.J. Rosansky and M.E. Brady, *Am. J. Kidney Dis.*, **7**, 502 (1986).

59. C.L. Low, K.R. Phelps, and G.R. Bailie, *Nephrol. Dial. Transplant*, **11**, 881 (1996).

60. P.D. Kroboth, K. Mitchum, and J.B. Purchett, *Am. J. Kidney Dis.*, **4**, 78 (1985).

61. C.L. Raehl, A.V. Moorthy, G.J. Beirne, and M.E. Pitterle, *Clin. Pharm.*, **4**, 669 (1985).

62. D.A. Sica, C. Yonce, R. Small, E. Cefali, A. Harford, and W. Poynor, *Int. J. Clin. Pharmacol. Therap. Toxicol*, **26**, 59 (1988).

63. T.A. Golper, J. Pulliam, and W.M. Bennett, *Arch. Intern. Med.*, **145**, 1651 (1985).

64. V.P. Kalashnikov and A.F. Mynka, *Farm. Zh.*, **4**, 32 (1984).

65. J. Sterling, S. Cox, and W.G. Haney, *J. Pharm. Sci.*, **63**, 1744 (1974).

66. B.D. Yurkevich, B.T. Ivanchenko, ***Sezda Farm.***, 3rd edn., 114-115 (1977), Edited by I.F. Urvantsev, Minsk. Gos. Med. Inst.; Minsk, USSR.

67. ***American Chemical Society Specification 1968***, 4th Edition, American Chemical Society Publications, Washington D.C., 1968.

68. H. Kadin, *J. Pharm. Sci.,* **63**, 919 (1974).

69. Y. Asahi, M. Takana, and K. Mima, *Bunseki Kagaku,* **35**, 147 (1986).

70. S. Zommer-Urbanska, and J. Urbanska, *Pharmazie,* **40**, 419 (1985).

71. R. Lewandowski, T. Sokalski, and A. Hulanicki, *Clin. Chem.* **35**, 2146 (1989).

72. P.N. Catania and J.C. King, *Am. J. Hosp. Pharm.*, **29**, 141 (1972).

73. J.C. Hamm, *J. Assoc. Off. Anal. Chem.*, **59**, 807 (1976).

74. A. Terlingo, M. Sarkar, E. Sypniewski, and T. Karnes, *ASHP Midyear Clinical Meeting,* **24**, P-109 E (1989).

75. A.F. Mynka, A.A. Murav'ev, and M.L. Lyuta, *Farm. Zh,* **3,** 42 (1980).

76. F. Plavsic, *Period. Biol.*, **82**, 289 (1980).

77. N.A. Klitgaard, *Arch. Pharm. Chem. Sci. Ed.,* **4**, 61 (1976).

78. H.S.I. Tan, J.M. Doepker, and S.K. Chia, *Asian J. Pharm. Sci.,* **1**, 51 (1979).

79. J.E. Whitaker and A.M. Hoyt, Jun., *J. Pharm. Sci.,* **73**, 1184 (1984).

80. C.S.P. Sastry, P.L. Kumari, and B.G. Rao, *Chem. Anal.,* **30**, 461 (1985).

81. S.A. Filipeva, L.M. Strelets, V.V. Petrenko, and V.P. Buryak, *Farm. Zh.*, **3**, 73 (1988).

82. S.A. Filipeva, L.M. Strelets, V.V. Petrenko and V.P. Buryak, *Zh. Anal. Khim.*, **44**, 131 (1989).

83. G.R. Rao, A.B. Avadhanulu, and D.K. Vasta, *East Pharm.*, **33**, 147 (1990).

84. C.S.P. Sastry, T.T. Rao, and A. Sailaja, *Talanta*, **38**, 1057 (1991).

85. A.M.I. Mohamed, H.Y. Hassan, H.A. Mohamed, and S.A. Hussein, *J. Pharm. Biomed. Anal.*, **9**, 525 (1991).

86. D.S. Sitar, D.N. Graham, R.E. Rangno, L.R. Dusfresne, and R.I. Ogilvie *Clin. Chem.*, **22**, 379 (1976).

87. T. Gurkan, *Istanbul Univ. Eczacilik Fak. Mecm.*, **18**, 116 (1982).

88. A.F. Minka, I.I. Kopiichuk, V.I. Shkadova, and V.P. Kalashnikov, *Pharm. Zh.*, **2**, 37 (1984).

89. L.H. Peters, *Clin. Chem.*, **22**, 1241 (1976).

90. H.S.I. Tan and C. Beiser, *J. Pharm. Sci.*, **64,** 1207 (1975).

91. H.S.I. Tan S.S. McEnaney, and A.C. Glasser, *Microchem. J.*, **24**, 395 (1979).

92. B.E. Pape, *J. Anal. Toxicol.*, **6**, 44 (1982).

93. J.M. Sterling and W.G. Haney, *J. Pharm. Sci.*, **63**, 1448 (1974).

94. K. Frislid, J.E. Bredesen, and P.K. Lunde, *Clin. Chem.*, **21**, 1180 (1975).

95. P.K. Ambler and J.R. Masarei, *Clin. Chem. Acta*, **70**, 379 (1976).

96. W.J. Long, R.C. Norin, and S.Y. Su, *Anal. Chem.*, **57**, 2873 (1985).

97. S.M. Sultan and F.E.O. Soliman, *Talanta*, **40**, 623 (1993).

98. M. Silver, *Eur. Clin. Lab.*, **26** (1990).

99. K. Nikolic, S. Vladimirov, D. Zivanov-Stakic, and K. Velasevic, *Acta Pol. Pharm.*, **44**, 438 (1987).

100. E. Bishop and W. Hussein, *Analyst*, **109**, 65 (1984).

101. J.F. Wheeler, C.E. Lunte, H. Zimmer, and W.R. Heineman, *J. Pharm. Biomed. Anal*, **8**, 143 (1990).

102. T. Katsu, K. Furuno, S. Yamashita, and Y. Gomita, *Anal. Chem. Acta*, **312**, 35 (1995).

103. J.T.R. Owen, R. Sithiraks, and F.A. Underwood, *J. Assoc. Off. Anal. Chem.*, **55**, 1171 (1972).

104. S. Sarre, K. Van-Belle, I. Smolders, G. Krieken, and Y. Michotte, *J. Pharm. Biomed. Anal.*, **10**, 735 (1992).

105. A.S. Curry and H. Powell, *Nature*, **173**, 1143 (1954).

106. H.V. Street, *Acta Pharmacologica et Toxicologia*, **19**, 312, 325 (1962).

107. H.V. Street, *J. Pharm. Pharmacol.*, **14**, 56 (1962).

108. H.V. Street, *J. Fones. Sci.*, **2**, 118 (1962).

109. I. Sunshine, *J. Clin. Pathol.*, **40**, 576 (1963).

110. R.N. Gupta, F. Eng, and D. Lewis, *Anal. Chem.* **50**, 197 (1978).

111. B. Kark, N Sistovaris, and A. Keller, *J. Chromatogr. Biomed. Appl.*, **277**, 261 (1983).

112. N. Sistovaris, *Chromatogr.*, **3**, 17 (1983).

113. G. Kufner and H. Schlegel, *J. Chromatogr.*, **169**, 141 (1979).

114. N. Sistovaris, A. Keller, and B. Kark, *J. Chromatogr. Biomed. Appl.*, **277**, 261 (1983).

115. G. Malikin, S. Lam, and A. Karmen, *Chromatographia,* **18**, 253 (1984).

116. K. Dross, C. Sonntag, and R. Mannhold, *J. Chromatogr.,* **639**, 287 (1993).

117. A.H. Stead, R. Gill, T. Wright, G.P. Gibbs, and A.C. Moffat, *Analyst*, **107**, 1106 (1982).

118. B. Wesley-Hadzija and A.M. Mattocks, *J. Chromatogr.*, **143**, 307 (1977).

119. K.Y. Lee, D. Nurok, and A. Zlatkis, *J. Chromatogr.*, **158**, 403 (1978).

120. K.J. Simons and R.H. Levy, *J. Pharm. Sci.,* **64**, 1967 (1975).

121. K.J. Simons, R.H. Levy, R.E. Cutler, G.T. Christopher, and A. Lindner, *Res. Commun. Chem. Pathol. Pharmacol.,* **11**, 173 (1975).

122. J.E. Carter, J.S. Dutcher, D.P. Carney, L.R. Klein, L.A. Black, and P.W. Erhardt, *J. Pharm. Sci.,* **69**, 1439 (1980).

123. K.M. Kessler, P. Ho-Tung, B. Steele, J. Silver, A. Pickoff, S. Narayanan, and R.J. Myerburg, *Clin. Chem.*, **28**, 1187 (1982).

124. A. Yamaji, K. Kataoka, M. Oishi, N. Kanamori, E. Hiraoka, and M. Mishima, *J. Chromatogr. Biomed. Appl.,* **59**, 143 (1987).

125. T.M. Ludden, D. Lalka, M.G. Wyman, B.N. Goldreyer, K.D. Haegele, D.T. Brooks, I. Davila, and J.E. Wallace, *J. Pharm. Sci.,* **67**, 371 (1978).

126. R.E. Andrey and A.C. Moffat, *J. Chromatogr.*, **220**, 195 (1981).

127. R.J. Flanagan and D.J. Berry, *J. Chromatogr.*, **131**, 131 (1977).

128. I.L. Zhuravleva, M.B. Terenina, R.V. Golovnya, and M.A. Filimonova, *Khim. Farm. Zh.* **5**, 58 (1993).

129. E.S.P. Bouvier, D.M. Martin, P.C. Iraneta, M. Capparella, Y.F. Cheng, and D.J. Phillips. *LC-GC Int*, **10**, 577, 582, 585 (1997).

130. E. Lessard, A. Fortin, A. Coquet, P.M. Belanger, B.A. Hamelin, and J. Turgeon, *J. Chromatogr. Sci.*, **36**, 49 (1998).

131. Chrompack Application Note, 959, Sept 1999, p. 1. See http://www.chrompack.com.

132. D.V. McCalley, *J. Chromatogr.*, **844**, 23 (1999).

133. C.A. Carnes and J.D. Coyle, *Pharmaco Therapy*, **12**, 40 (1992).

134. R.R. Bridges and T.A. Jennison, *J. Anal. Toxicol.*, **7**, 65 (1983).

135. J.F. Wesley and F.D. Lasky, *Clin. Biochem.*, **15**, 284 (1982).

136. F.M. Stearns, *Clin. Chem.,* **27**, 2064 (1981).

137. W.C. Griffiths, P. Dextraze, M. Hayes, J. Mitchell, and I. Diamond, *Clin. Toxicol.*, **16**, 51 (1980).

138. F.L. Vandemark, *Chromatogr. Newslett.,* **8**, 57 (1980).

139. C. Gallaher, G.L. Henderson, R.I. Low, E.A. Amsterdam, and D.T. Mason, *J. Chromatogr. Biomed. Appl.*, **7**, 490 (1980).

140. P. Gyselinck, R. Van-Severen, P. Braeckman, and E. Schacht, *J. Pharm. Belg.,* **36**, 200 (1981).

141. T.I. Ruo, J.P. Thenot, G.P. Stec, and A.J. Atkinson, Jun., *Therap. Drug. Monit.,* **3**, 231 (1981).

142. R.L. Nation, M.G. Lee, S.M. Huang, and W.L. Chiou, *J. Pharm. Sci.,* **68**, 532 (1979).

143. M.A. Gadalla, G.W. Peng, and W.L. Chiou, *J. Pharm. Sci.*, **67**, 869 (1978).

144. A.G. Butterfield, J.K. Cooper, and K.K. Midha, *J. Pharm. Sci.*, **67**, 839 (1978).

145. P.O. Lagerstrom and B.A. Persson, *J. Chromatogr.*, **149**, 331 (1978).

146. N. Verbiese-Genard, M. Hanocq, M. Van-Damme, and L. Molle, *Int. J. Pharm.*, **2**, 155 (1979).

147. I. Jane, A. McKinnon, and R.J. Flanagan, *J. Chromatogr.*, **323**, 191 (1985).

148. S.Q. Zhang, G.C. Olugo-Edege, Y.P. Zhu, F.F. Mao, and D.K. An, *Acta Pharm. Sinica*, **23**, 430 (1988).

149. E. VasBinder and T. Annesley, *Biomed. Chromatogr.*, **5**, 19 (1991).

150. R. Verbesselt, T.B. Tjandramaga, and P.J. De-Schepper, *Therap. Drug Monit.*, **13**, 157 (1991).

151. M. Kim and J.T. Stewart, *Microchim. Acta.*, **3**, 221 (1990).

152. H. Imai and G. Tamai, *Biomed. Chromatogr.*, **3**, 192 (1989).

153. D.J. Gisch, R. Ludwig, and L.A. Witting, *J. High Resolut. Chromatogr.*, **12**, 409 (1989).

154. J.V. Posluszny and R. Weinberger, *Anal. Chem.*, **60**, 1953 (1988).

155. G. Tamai, H. Yoshida, and H. Imai, *J. Chromatogr. Biomed. Appl.* **67**, 155 (1987).

156. J.D. Coyle, J.J. Mackichan, H. Boudoulas, and J.J. Lima, *J. Pharm. Sci.*, **76**, 402 (1987).

157. S.H.Y. Wong, N. Marzouk, O. Aziz, and S. Sheeran, *J. Liq. Chromatogr.*, **10**, 491 (1987).

158. D. Raphanaud, M. Borensztejn, J.P. Dupeyron, and F. Guyon, *Therap. Drug Monit.*, **8**, 365 (1986).

159. H.F. Proelss and T.B. Townsend, *Clin. Chem.*, **32**, 1311 (1986).

160. T. Annesley, K. Matz, R. Davenport, and D. Giacherio, *Res. Commun. Chem. Pathol. Pharmacol.*, **51**, 173 (1986).

161. H. Yoshida, I. Morita, G. Tamai, T. Masujima, T. Tsuru, and N. Takai, *Chromatographia*, **19**, 466 (1984).

162. L.E. Kholodov, A.V. Sokolov, A.A. Dragunov, R.V. Makharadze, and I.F. Tishchenkova, *Khim. Farm. Zh.*, **17**, 868 (1983).

163. G. Tamai, H. Yoshida, and H. Imai, *J. Chromatogr.*, **423**, 155 (1987).

164. H.J. Willems, A. Van-der-Horst, P.N. De-Goede, and G.J. Haakmeester, *Pharm. Weekbl., Sci.*, **7**, 150 (1985).

165. C.P. Patel, *Therap. Drug Monit.*, **5**, 235 (1983).

166. C.N. Ou and V.L. Frawley, *Clin. Chem.*, **28**, 2157 (1982).

167. J.F. Wesley, and F.D. Lasky, *Clin. Biochem.*, **14**, 113 (1981).

168. P.M. Kabra, S.H. Chen, and L.J. Marton, *Therap. Drug Monit.*, **3**, 91 (1981).

169. A.J. Quattrone and R.S. Putnam, *J. Anal. Toxicol.*, **5**, 101 (1981).

170. C.M. Lai, B.L. Kamath, Z.M. Look, and A. Yacobi, *J. Pharm. Sci.*, **69**, 982 (1980).

171. E.S.P. Bouvier, D.M. Martin, P.C. Iraneta, M. Capparella, Y.F. Cheng, and D.J. Phillips, *LC-GC*, **15**, 152, 154, 158 (1997).

172. K. Shahdeo, C. March, and H.T. Karnes, *Anal. Chem.*, **69**, 4278 (1997).

173. Z.K. Shihabi, J. Scaro, and R.D. Dyer, *J. Liq. Chromatogr.*, **11**, 2391 (1988).

174. R.F. Adams, F.L. Vandemark, and G. Schmidt, *Clin. Chem. Acta*, **69**, 515 (1976).

175. K. Carr, R.L. Woosley, and J.A. Oates, *J. Chromatogr.*, **129**, 363 (1976).

176. L.R. Shukur, J.L. Powers, R.A. Marques, M.E. Winter, and W. Sadee, *Clin. Chem.*, **23**, 636 (1977).

177. R.M. Rocco, D.C. Abbott, R.W. Giese, and B.L. Karger, *Clin. Chem.*, **23**, 705 (1977).

178. O.H. Weddle and W.D. Mason, *J. Pharm. Sci.*, **66**, 874 (1977).

179. J.S. Dutcher and J.M. Strong, *Clin. Chem.*, **23**, 1318 (1977).

180. J.J. Mackichan, J.D. Coyle, B.J. Shields, H. Boudoulas, and J.J. Lima, *Clin. Chem.*, **30**, 768 (1984).

181. S. Lam, J. Al-Razi, and A. Karmen, *Chromatogr. Newslett.*, **7**, 11 (1979).

182. N.C. Gillott, M.R. Euerby, C.M. Johnson, D.A. Barrett, and P.N. Shaw, *Anal. Comm.* **35**, 217 (1998).

183. M.R. Euerby, D. Gilligan, C.M. Johnson, S.C.P. Roulin, P. Myers, and K.D. Bartle. *J. Microcolumn. Sep.*, **9**, 373 (1997).

184. N.C. Gillott, M.R. Euerby, C.M. Johnson, D.A. Barrett, and P.N. Shaw. *Chromatographia*, **51**, 167 (2000).

185. L.V. Pesakhovych, *Farm. Zh.*, **6**, 64 (1983).

186. C.V. Thomas, A.C. Cater, and J.J. Wheeler, *J. Liq. Chromatogr.*, **16**, 1903 (1993).

187. G. Vargas, J. Havel, and E. Hadasova, *J. Chromatogr.*, **772**, 271 (1997).

188. Z.K. Shihabi, *Electrophoresis,* **19**, 3008 (1998).

189. F.J. DeLuccia, M. Arunyanart, and L.J.C. Love, *Anal. Chem.,* **57**, 1564 (1985).

190. S.R. Needham and P.R. Brown, *J. Pharm. Biomed. Anal.*, **23**, 597 (2000).

191. P. Mojaverian and G.D. Chase, *J. Pharm. Sci.*, **69**, 721 (1980).

192. P.R. Bach and J.W. Larsen, *Clin Chem.*, **26**, 652 (1980).

193. V.M. Haver, N. Audino, S. Burris, and M. Nelson, *Clin Chem.*, **35**, 138 (1989).

194. C.B. Walberg and S.H. Wan, *Therap. Drug. Monit,* **1**, 47 (1979).

195. P. Spitzbarth, *Pharm. Weekbl.*, **117**, 479 (1982).

196. W. Shaw and J. McHan, *Therap. Drug Monit.* **3**, 185 (1981).

197. G.V. Zvaigzne, D.A. Brogan, and L.H. Bernstein, *Clin. Chem.*, **32**, 437 (1986).

198. P.R. Henry and R.A. Dhruv., *Clin. Chem.*, **34**, 957 (1988).

199. M.H.H. Al-Hakiem, D.S. Smith, and J. Landon, *J. Immunoassay*, **3**, 91 (1982).

200. P.K. Sonsalla, R.R. Bridges, T.A. Jennison, and C.M. Smith, *J. Anal. Toxicol.*, **9**, 152 (1985).

201. Y.W. Lam and C.E. Watkins, *Clin. Pharm.*, **12**, 49 (1993).

CUMULATIVE INDEX

Bold numerals refer to volume numbers.